ROBOTICS RESEARCH AND TECHNOLOGY

ROBOTIC AUTONOMY AND CONTROL: ARMY RESEARCH

ROBOTICS RESEARCH AND TECHNOLOGY

Additional books in this series can be found on Nova's website
under the Series tab.

Additional E-books in this series can be found on Nova's website
under the E-books tab.

ROBOTICS RESEARCH AND TECHNOLOGY

ROBOTIC AUTONOMY AND CONTROL: ARMY RESEARCH

TAYLOR C. MUELLER
AND
MELISSA E. BYNES
EDITORS

Nova Science Publishers, Inc.
New York

Copyright © 2012 by Nova Science Publishers, Inc.

All rights reserved. No part of this book may be reproduced, stored in a retrieval system or transmitted in any form or by any means: electronic, electrostatic, magnetic, tape, mechanical photocopying, recording or otherwise without the written permission of the Publisher.

For permission to use material from this book please contact us:
Telephone 631-231-7269; Fax 631-231-8175
Web Site: http://www.novapublishers.com

NOTICE TO THE READER

The Publisher has taken reasonable care in the preparation of this book, but makes no expressed or implied warranty of any kind and assumes no responsibility for any errors or omissions. No liability is assumed for incidental or consequential damages in connection with or arising out of information contained in this book. The Publisher shall not be liable for any special, consequential, or exemplary damages resulting, in whole or in part, from the readers' use of, or reliance upon, this material. Any parts of this book based on government reports are so indicated and copyright is claimed for those parts to the extent applicable to compilations of such works.

Independent verification should be sought for any data, advice or recommendations contained in this book. In addition, no responsibility is assumed by the publisher for any injury and/or damage to persons or property arising from any methods, products, instructions, ideas or otherwise contained in this publication.

This publication is designed to provide accurate and authoritative information with regard to the subject matter covered herein. It is sold with the clear understanding that the Publisher is not engaged in rendering legal or any other professional services. If legal or any other expert assistance is required, the services of a competent person should be sought. FROM A DECLARATION OF PARTICIPANTS JOINTLY ADOPTED BY A COMMITTEE OF THE AMERICAN BAR ASSOCIATION AND A COMMITTEE OF PUBLISHERS.

Additional color graphics may be available in the e-book version of this book.

Library of Congress Cataloging-in-Publication Data

Robotic autonomy and control : Army research / editors, Taylor C. Mueller and Melissa E. Bynes.
    p. cm.
  Includes index.
  ISBN 978-1-62100-605-3 (hardcover)
  1. Military robots--United States. 2. Robots--Control systems. 3. Robotics--Military applications--United States. 4. United States--Armed Forces--Robots. I. Mueller, Taylor C. II. Bynes, Melissa E.
  UG479.R63 2011
  629.8'95--dc23

2011037396

Published by Nova Science Publishers, Inc. † New York

CONTENTS

Preface		**vii**
Chapter 1	A Platform for Developing Autonomy echnologies for Small Military Robots *Gary Haas, Jason Owens and Jim Spangler*	**1**
Chapter 2	Scalability of Robotic Controllers: Effects of Progressive Levels of Autonomy on Robotic Reconnaissance Tasks *Rodger A. Pettitt, Elizabeth S. Redden,* *Estrellina Pacis and Christian B. Carstens*	**31**
Chapter 3	Intuitive Speech-Based Robotic Control *Elizabeth S. Redden, Christian B. Carstens* *and Rodger A. Pettitt*	**83**
Index		**125**

PREFACE

This book explores robotic autonomy and control with a focus on the U.S. Army Research Laboratory's initiation of research targeting robotics technologies for asymmetric warfare in urban terrain. The vision for this tactical domain calls for small robots suited to maneuver indoors as well as outdoors, perhaps sharing space with troops, with sensors to perceive the immediate surroundings, and sufficient intelligence to enable unsupervised operation. Also discussed is the scalability of robotic controllers and the effects of progressive levels of autonomy on robotic reconnaissance tasks and intuitive speech-based robotic control.

Chapter 1- In late 2007, the U.S. Army Research Laboratory (ARL) initiated a research thrust targeting robotics technologies for asymmetric warfare in urban terrain. The vision for this tactical domain calls for small robots suited to maneuver indoors as well as outdoors, perhaps sharing space with troops, with sensors to perceive the immediate surroundings, and sufficient intelligence to enable (at least short periods of) unsupervised operation. Such a robot must embody substantial autonomy, e.g., be able to "see" and "understand" its environment so that it can perform its function with minimum burden on the soldiers it supports. It requires sensors capable of detecting the immediate surroundings with high fidelity and richness, and powerful onboard computing systems. At the inception of the new thrust, such capabilities were unavailable in research robots currently in ARL's robotics labs. As a first step toward exploring this new mission space, scientists and engineers at the Vehicle Technology Directorate's Unmanned Vehicle Technologies Division (UVTD) set out to create the capability.

Chapter 2- Maes proposed a continuum of robot control that ranges from teleoperation to full autonomy. Robots can be placed on this continuum based on the level of human-robot interaction that is required. Teleoperation is the lowest level of automation on the continuum because it requires the most intervention from the robot operator. A teleoperated robot is totally under the control of the operator who uses a joystick or other control device to command the robot. This requires constant interaction between the robot and the operator. Teleoperation is not the ideal solution for all situations. Many teleoperation tasks are repetitive and boring and the work requires constant attention by the operator.

Chapter 3- Two previous studies conducted by the U.S. Army Research Laboratory's (ARL) Human Research and Engineering Directorate (HRED) indicated that the degree of effectiveness of speech-based control may be dependent upon the task being performed. The goal of the present research was to examine the effectiveness of speech control for the specific tasks used in robotic reconnaissance missions.

In: Robotic Autonomy and Control
Editors: T.C. Mueller and M.E. Bynes

ISBN: 978-1-62100-605-3
© 2012 Nova Science Publishers, Inc.

Chapter 1

A PLATFORM FOR DEVELOPING AUTONOMY TECHNOLOGIES FOR SMALL MILITARY ROBOTS[*]

Gary Haas, Jason Owens and Jim Spangler

1. OBJECTIVE AND PROGRAM BACKGROUND[1]

In late 2007, the U.S. Army Research Laboratory (ARL) initiated a research thrust targeting robotics technologies for asymmetric warfare in urban terrain. The vision for this tactical domain calls for small robots suited to maneuver indoors as well as outdoors, perhaps sharing space with troops, with sensors to perceive the immediate surroundings, and sufficient intelligence to enable (at least short periods of) unsupervised operation. Such a robot must embody substantial autonomy, e.g., be able to "see" and "understand" its environment so that it can perform its function with minimum burden on the soldiers it supports. It requires sensors capable of detecting the immediate surroundings with high fidelity and richness, and powerful onboard computing systems. At the inception of the new thrust, such capabilities were unavailable in research robots currently in ARL's robotics labs. As a first step toward

---

[*] This is an edited, reformatted and augmented version of an Army Research Laboratory publication, ARL-MR-709, dated December 2008.

exploring this new mission space, scientists and engineers at the Vehicle Technology Directorate's Unmanned Vehicle Technologies Division (UVTD) set out to create the capability.

A test bed for developing autonomy technologies, at least early in the program, can be based on commercial sensing technologies and a mobility platform of limited performance. The lab had in its inventory 8-year-old ATRV Jr. research robots once built by Real World Interface, Inc. (RWII). At the time of acquisition, these robots were quite advanced and offered skid-steer wheeled mobility, global positioning system (GPS) and electronic compass for navigation, ladar (a portmanteau of laser radar and often used interchangeably with lidar) and sonar sensors for obstacle detection, and a software development environment based on linked server modules. These robots were at the core of in-house robotics research at ARL. By 2007, the ATRVs were well worn. RWII (renamed iRobot Corporation) had discontinued production and support, and the Pentium III processor and Red Hat 6.2 operating system at the core of the robotics had been superceded by several generations.

Research supporting ARL's new robotics thrust calls for research robots similar in scale to the old ATRVs and with power and payload to support quantities of sensors and computing. New research robots have become available from several vendors, but, in general, the function of these products is not substantially different from that of the old ATRVs. The decision was made to renovate the old robots rather than invest in new research robots. This report describes the upgrade of the ATRV robot for its new role.

2. SYSTEM DESCRIPTION

The stock ATRV robot is a 25-in-long ,24.5-in-wide ,21-in-high 110-lb vehicle. Two deep-draw, gel-cell, lead-acid batteries power a pair of servo motors that drive its four 12-in tires in skid-steer fashion by means of toothed belts. A sturdy rectangular sheet-metal chassis houses the internals and supports the mounting rails front and rear and on the deck.

Of the rest of the original major components, only the pan-tilt unit remains. The rest have been replaced by functional equivalents and supplemented with functional extrapolations. Figure 1 depicts the components of the upgrade at a block diagram level. Details of the upgrade follow.

A Platform for Developing Autonomy Technologies ...

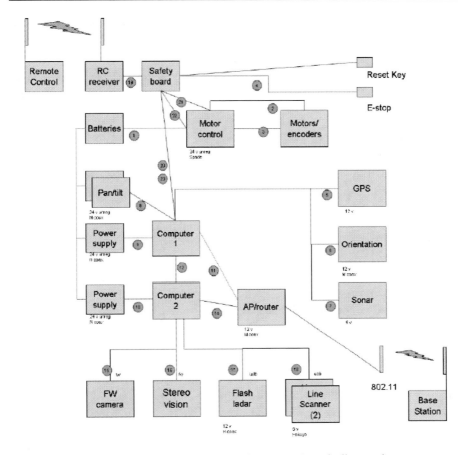

Figure 1. Major components of the upgraded ATRV. Gray indicates the component is off-board, linked by radio.

3. UPGRADE

3.1. Hardware

3.1.1. Computational Hardware

Increasing the computational capacity was a requirement for the ATRV upgrades. In order to accommodate the planned sensor-processing research, the computers needed to represent the latest technology available off the shelf.

In addition, they needed to be small and power-efficient but possess sufficient processing capability to minimize the processing bottleneck. The current trend in the hardware and software community is to leverage horizontal scale (more processing elements on a die), more so than clock speed. Thus the new computational hardware design takes full advantage of the space and processors available.

3.1.1.1. Central Processing Units (CPUs)

A Mini-ITX (a motherboard format popularized by Via Technologies, Inc. [1]) was located that supported dual- and quad-core Intel processors. The only one available at the time was the Commell Core 2 Quad[2] with dual gigabit ethernet ports, six USB 2.0 ports, an 8-bit digital general purpose input/output[3] port, and Serial Advanced Technology Attachment.[4] The ATRV Jr. has enough room for two reasonably sized mini-ITX cases, so the computers are relatively modular and easily replaceable. Each computer is equipped with 4 GB of RAM (although the computer system bus allows access to only 3 GB). A gigabit ethernet switch connects the two machines together and essentially yields a miniature computing cluster with eight processor cores.

3.1.1.2. Storage

Storage is provided by a single 16-GB solid-state drive for each CPU. Solid-state drives are used to increase system performance (relative to standard rotating platter 4200 RPM laptop drives) and increase system reliability with respect to vibration issues. While the storage size is small compared to today's large drives, it is more than enough to hold an hour's worth of raw data from the primary sensors and can be easily expanded with larger sizes if the need arises.

3.1.1.3. Network

As mentioned previously, the computers interface with each other through the network switch and with external machines (i.e., control stations, logging/debugging systems, development systems) through either a 100-Mbs wired or 54-Mbs wireless interface provided by an Alfa Network's AWAP608 wireless access point [2]).

3.1.1.4. Sensor Interfaces

The ATRV Jr. has a variety of onboard sensors that utilize several connection interfaces on the CPU: serial, USB 2.0, and IEEE 1394.[5] Many of the serial devices are connected through RS-232[6]-to-USB adapters, while others (like the motor controller) are connected directly to the motherboard. Several perception sensors on the ATRV Jr. use USB and IEEE 1394 directly.

3.1.2. Sensors for Reaction and World Modeling

The purpose of sensors on a mobile robot is to create an analog of the nearby environment in data structures used by the computer programs that control the robot. These data structures are collectively termed the "world model." While some robots simply react to sensor inputs to alter some behavior ("obstacle ahead, turn left"), the objective of this project is to enable the robot to sense the geometry of its environs, store the sensed elements in data structures based on a self-constructed local map, and plan its behaviors based on the map. The map will have different layers, populated with geometric elements (extracted and abstracted from its geometry sensors), spectral elements (extracted from its cameras), elements fused from the sensor-derived elements, and iconic elements (extrapolated from the sensor-derived elements and filtered based on a mission-based context).

The universe of sensors appropriate for small robots is not a large one. Sensors considered for the robot are described further in this section, and those selected for the upgrade are pictured in figure 2.

3.1.2.1. Video Camera

Video cameras are widely available and relatively inexpensive. Imagery is dense (high pixel count) but only spectral in nature. The information content of the image is rich but lacks the immediate geometric significance needed for safe mobility. Significant processing is necessary to convert a stream of spectral images to the geometric world model needed. However, given a geometry by some other sensor, the richness of the video imagery can be overlaid. Video imagery is also the most easily interpreted sensor mode for a human. Augmented by feature-tracking software, the video sensor can provide a direction reference and can support algorithms such as direction-only simultaneous localization and mapping.

Given the widespread availability of video cameras, there are a number of parameters that can be used as a selection criterion. Field of view, a function of lens selection, is probably the most important (the wider the better). In general, the image resolution (number of pixels) is not important, as even the least capable cameras have sufficient resolution for the application (except for stereo vision, which will be treated separately). A key parameter is the ease of integration with a computer. While analog cameras today dominate the market, cameras with a built-in digital interface are more suitable for the application.

This is partly due to ease of interfacing, but the primary reason is that most digital video cameras have progressive scan technology, which is important when the camera is mounted on a moving platform. Firewire (IEEE 1394) cameras are the most common. For streaming applications, Universal Serial Bus (USB) has little to offer over Firewire, which was designed with video applications in mind. Both offer 400-Mbs rates, adequate for video graphics array[7] (VGA)-quality resolution or a little more. The newer IEEE 1394b, at 800 Mbs, is not yet widely available but will become so.

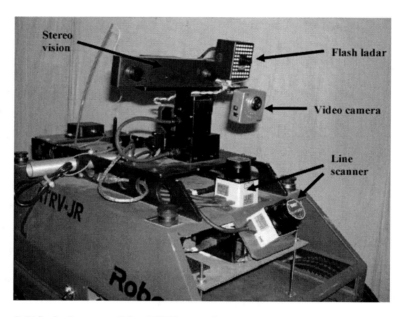

Figure 2. Principal sensor of the ATRV upgrade.

A representative video camera is the Unibrain Fire-i, priced at around $120 for VGA-resolution imagery with a plastic case and glass lens [3]. A 4.3-mm focal length f 2.0 lens was selected, specified to deliver VGA-resolution imagery at 30 Hz, covering a horizontal field of view of 42.25°. A number of vendors offer competing products; this selection was based on low price for adequate performance.

3.1.2.2. Near-Range Geometry Sensing

For a number of years the only sensor capable of detecting the geometry of the environment, in a package of suitable size and cost for use on a small robot, was a sonar sensor such as the Polaroid product of the early 1980s.

Modern sonar sensors for robotics differentiate between empty and occupied volumes in a cone subtending as little as 15°, so the spatial resolution perpendicular to the cone is limited, and the "occupied" region of the cone can be a tiny fraction of the cross section of the cone at the detected range [4]. Sonar units work well enough for simple reactive obstacle detection, and several will be incorporated in the finished upgrade. However, a denser scan is necessary to be useful as a geometry sensor.

3.1.2.2.1. Line Scanner

A line scanner is one such sensor and is widely used in mobile robotics. A line scanner is a laser range finder, which is swept through an arc by a spinning mirror. The sensor detects the return from the laser and calculates distance from time of flight at discrete angular increments around the disk so described. A line scanner oriented so the plane of detected points is horizontal (e.g., the axis about which the mirror spins is vertical and the angle between the mirror and its axis is 45°) is useful in real-world terrain where objects of interest (walls, etc.) are also vertical, such as an indoor environment. This sort of sensor is insufficient for general obstacle detection and terrain mapping but can be used to generate a useful first approximation, as vertical terrain features tend to be the most salient.

A line-scanning unit built by the German company SICK AG was used on many of the mobile robots competing in the recent Defense Advanced Research Projects Agency Grand Challenges for autonomous unmanned ground vehicles. The SICK unit, however, is too large and heavy for this application. Instead, a device similar to the principal geometry sensor was used.

A smaller line-scanning device, the URG-04LX [5], is available from Hokuyo. This sensor works very much like the SICK scanner but is small (2 x 2 x 3 in) and lightweight (165 g). It sweeps an arc of 240° at a rate of 10 Hz, returning range measurements at intervals of 0.36°. This corresponds to approximately one data point per inch at the maximum range of 4 m and 683 data points every 100 ms to process to maintain real time. The URG sensor is mounted to the frame of the robot so that the plane of the measurements is horizontal and at the height of the robot. This is consistent with using the sensor to avoid right prismatic (cuboid) obstacles, and it also enables the mapping of indoor terrain, which is predominantly bounded by vertical planes.

A second line scanner is mounted at the front of the robot, directed at the ground ~1 m ahead of the robot. This scanner senses the terrain the robot is just about to drive onto. The horizontal line scanner receives no sensed data from the ground, so the second scanner is depended upon to assure that there is indeed ground to drive upon and that the terrain is smooth enough for the robot to traverse. Ideally, this sensor would look out 3 m ahead so there would be time to stop if, for example, the sensor detected the top step of a flight of stairs. The look-ahead distance may be changed as researchers gain experience with the system.

3.1.2.2.2. Imaging Ladar

More detail concerning the geometry of the environs is available from an imaging ladar sensor. Such a sensor acquires range data as a set of range vectors centered at a focal point and organized as an image, e.g., rows and columns of data points. Surveying ladars, such as those available from Riegl USA, Inc., provide high-resolution three-dimensional data at ranges over 100 m, but the range measurements are sequential and too slow for mobile applications. Ladars built specifically for mobility applications, such as the product built by General Dynamics Robotic Systems for the Army's Autonomous Navigation System, are substantially faster, but today's technology is too large and heavy for use on this small robot.

A recent technology known as "flash ladar" is based on camera technology, enabling fast data acquisition as well as light weight and compact size. Such a device illuminates the environment with light modulated at a known frequency and determines time of flight from the phase shift of the reflected energy incident on each pixel imager. Devices using this technology are available from PMD Technologies GmbH, Mesa Imaging AG, and possibly others.

The device selected for the ATRV sensor upgrade is the Mesa Imaging SwissRanger SR-3000 [6]. This sensor collects frames of range data 176 x 144 pixels at a rate of 30 Hz over a field of view of 47.5° x 39.6° (0.27° per pixel). Maximum range is advertised as 7.5 m, limited by the nonambiguity constraints of the measurement technique, but several papers in the literature indicate a shorter useful range. Range resolution is specified by the data sheet as 1% of range. A cursory evaluation of the sensor revealed a sensitivity to bright lights, resulting in washed-out regions of the image, which must be further investigated.

The SwissRanger is mounted on an existing pan-tilt unit on the deck of the ATRV. The pan will be used to compensate for the narrow fields of view of the various sensors mounted on the unit. The tilt axis will likely be set at a fixed look-down angle, which provides a "good" amount of information about the ground immediately ahead of the robot while not sacrificing too much information about overhanging objects.

3.1.2.2.3. Stereo Vision

There will be times when it is necessary to sense geometry at ranges greater than that provided by the active sensors. Computer-based stereo vision can provide range images at distances of tens of meters, but it has been seldom utilized outside the laboratory (and in planetary exploration). In part, this is because the sensors (conventional cameras) are inexpensive, but the computing to process the camera images into a range image was "do it yourself"—the phenomenon was well understood and algorithms were widely available, but there was no integrated stereo "system" delivering range images.

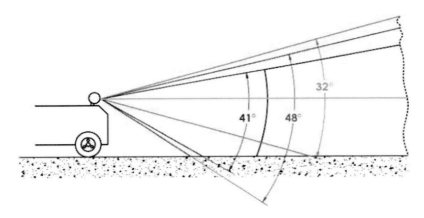

Figure 3. Sensor fields of view are shown here in side view, according to the following key: stereo in blue, color camera in green, flash ladar in red, horizontal line scanner in magenta.

The recent availability of "Stereo on a Chip," from Videre Design LLC, has changed the maximum range available from an active sensor [7]. A field programmable gated array packaged with the complementary metal oxide semiconductor video imagers computes disparity (a function of range) at each pixel of the VGA-resolution image at 30 Hz, reducing the workload of the host processor substantially. The 4.5-mm lens images a field of view of 59° horizontal x 41° vertical. As configured for UVTD's application, the range resolution at 8 m is computed by Videre Design's online calculator to be roughly 2.5 in; ranges closer than 0.5 m are unavailable. The stereo sensor is expected to deliver a dense sampling of a (possibly imprecise) range function, allowing ranges to be estimated beyond the ability of the ladar sensors. Reflectance values from the stereo system are also available on a frame-by-frame basis, enabling data integration and/or fusion.

The stereo sensor will be mounted on the pan-tilt unit near the SwissRanger so the region sampled by both sensors overlaps. The overlap among the fields of view of the various sensors, shown in figures 3 and 4, will be exploited in any way possible.

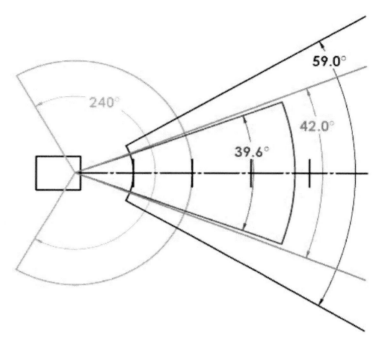

Figure 4. Sensor fields of view are shown here in top view, according to the following key: stereo in blue, color camera in green, flash ladar in red, horizontal line scanner in magenta. Black centerline shows crosshatches at 2-m intervals out to 10 m.

3.1.2.3. Inertial Reference

All the sensors described report their sensed data in the coordinate frame of the sensor itself. In order to integrate the information over time, it is necessary to transform the data to a common coordinate system, preferably a world-fixed coordinate system. The conventional way to do this is to sense the robot location from a sensor such as GPS and the robot orientation from a compass, and augment these sensors with time derivatives of each from an inertial reference sensor suite and odometry. In the case of this ATRV, the ideal operational area is where GPS is unavailable and where compass readings may be compromised (and possibly in unmapped regions). In this case, high-quality time derivatives of position and orientation are wanted because of the integrations required.

A 3DM-GX1 inertial reference sensor (IRS) from MicroStrain, Inc., [8] provides orientation and acceleration for the upgraded robot. Raw data from embedded accelerometer, gyros, and magnetometers are fused by the sensor itself. Alternative IRSs are available but were not considered since sensors from the MicroStrain product line are used on other division assets, and performance was deemed acceptable.

3.1.3. Power

Power for the robot as a whole was left unchanged from the original ATRV, that is, dual 12-V deep-draw batteries with off-board recharging. It remains to be seen whether the duration available will be sufficient for research missions.

Power for peripherals was shifted from the computer power supply of the original ATRV power architecture to a custom power distribution board supplying regulated 5 and 12 V through bussed terminal strips.

The CPU of the original ATRV computer was rated at ~30 W, while the CPU selected for the upgrade was rated at 125 W, so the computer power supply was upgraded as well. The onboard power supply, the M2-ATX 160W [9], was selected based on its tolerance of a wide range of input voltages (6–24 V) and its form factor, which corresponds to the dimensions of the case selected. The maximum input operating voltage, outlined in the specifications, is 24 V, which is marginal for a battery system consisting of two 12-V batteries. A more recent release, the M2- ATX-HV, claims an even higher maximum input voltage and would be a more conservative selection.

3.1.4. Motor Control Board

The motor control board was replaced as well, though the two wheelchair motors and gearboxes driving the ATRV's four wheels were retained. The AX3500 product from Roboteq [10]) was selected for the following reasons:

1) It is designed to run with 24-V rechargeable batteries.
2) It supplies maximum motor current of 40 A for each of two brush-type electric motors, well beyond the 11-A rating of the wheelchair motors.

3) Input to the motor control board through RS-232 is available. A translational velocity /rotational velocity command is native.

4) The board supports optical encoders, also part of the original ATRV motor suite. The encoders enable closed-loop velocity control, an essential element for control of a skid-steer vehicle where soil resistance is uncertain. In addition, the encoder counts can be monitored through the serial link, providing an odometry function.

5) It has two distinct safety shutoff modes, including one which can be easily asserted from a remote radio control (RC) controller.

While each of the elements listed are essential to the application, a number of other features bring added value to the control board, notably the ability to control the robot from an RC remote control. This capability is very useful in logistic operations, such as maneuvering the robot into its parking place at the end of the work day.

3.1.5. Safety Circuitry

The original ATRV was equipped with four e-stop mushroom buttons on the robot. To stop the robot in case of a software failure, it was necessary to approach the robot and push one of the buttons. One of the goals of the upgrade was to increase the safety of the robot by enabling a software-independent kill capability from a distance, so the robot can be brought to a halt without jeopardizing the operator.

Using a safety mode suggested by the manufacturer of the motor control board, there are now three means of stopping a runaway robot. The first is by means of the e-stop buttons on the robot. These cause an e-stop input on the motor control board to be activated. The second is a red e-stop button on the remote control. Pressing this button actuates the same input through one channel of the RC radio. Both require a reset from a key-switch on the robot body before motor control is restored.

The third mode stops the robot by a new mechanism. The motor controller can accept commands from either the serial link connected to the onboard computer or from the RC receiver linked by a dedicated radio channel to the remote control. The remote control determines which signals reach the control board input by means of an electromechanical relay.

The default control is from the remote control, and it must be ceded to the computer by a manual switch on the remote control. If the program running on the computer is judged by the operator to be in dangerous error, the operator can throw the switch on the remote control, which seizes control from the computer and returns it to the remote control in the hands of the operator. This safety paradigm is similar to that used on UVTD unmanned air assets. Figure 5 depicts the operator's remote operation and safety control. Figures 6 and 7 illustrate status monitors, allowing the operator to confirm elements of the robot control state.

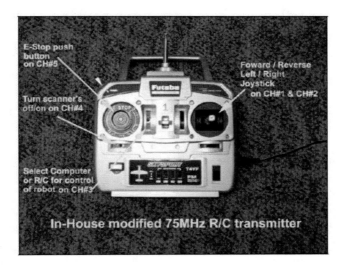

Figure 5. The remote control provides the operator the ability to stop the robot from a safe distance in case of emergency and operate it manually with the joystick.

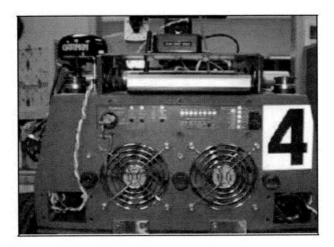

Figure 6. Display panel at ATRV rear.

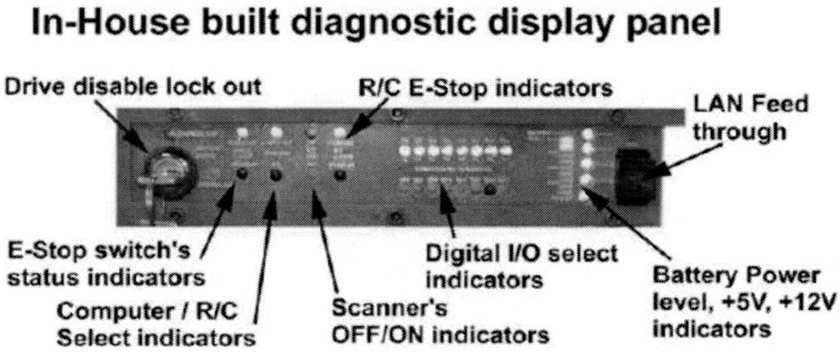

In-House built diagnostic display panel

Figure 7. The display panel shows the status of a number of internal states.

3.2. Robot Software

Since the original software provided by RWII for the ATRV Jr. was proprietary, replacing the computers meant replacing the operating system and supporting software (including the device drivers). Thus the decision was made to utilize the Open Source robotics package Player (from the Player/Stage project [11]), which provides a convenient hardware abstraction layer and a multitude of popular robotics device drivers. Player is also used by numerous academic institutions with robotics programs including the University of Pennsylvania, Georgia Institute of Technology, and the University of Southern California [12].

3.2.1. Organization

The software on the robot computers is divided into three layers: base, core, and brain (see figure 8).

The base layer contains relatively static library code and the core operating system facilities that are common across all machines, including a properly configured kernel for the CPU architecture. Debian [13] GNU/Linux is the operating system of choice, primarily to gain the benefit of its package management system Advanced Packaging Tool (APT) as well as the ease of configuring a custom system using the very useful debootstrap tool.

The core layer is an abstraction layer providing higher-level functionality to the layer above it. It contains the Player system and any shared libraries required for robotic development, including the agent architecture under development (see section 3.4). The ATRV Jr. currently under development has a custom driver plug-in for Player that was developed for the Roboteq controller, as well as the libraries required for the SR3000 Flash Ladar. In general, more volatile libraries (rapidly changing open source and internal packages) will reside in this layer.

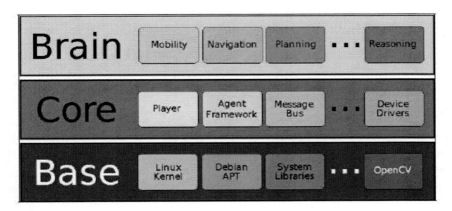

Figure 8. High-level software organization. The smaller items in each level indicate the types of components in each layer.

The brain layer is the effective application layer where agents and high-level behaviors can be implemented. Thus it will contain the custom scripts, executables, and data that compose the actual behavior of the robot. Currently, the brain layer is not implemented in the upgrade; however, work is underway to address that problem (section 3.4).

3.2.2. Operating System

The base operating system was constructed to be relatively small and boot fast. Debian GNU/Linux, however, provides an installation script that downloads a minimal Debian operating system without the Linux kernel. The minimal system includes only a small subset of a typical GNU/Linux distribution, so scripts were created to add in additional software readily available from the Debian software repository.

The omission of the Linux kernel from the minimal system is deliberate; more than likely, someone building a custom distribution will want to custom configure a kernel for specific hardware, as is the case for this project. Thus the scripts choose among several custom kernel configurations based on a given keyword, build the kernel, add it to the system base directory, and build a disk image that can be copied directly to the robot's hard disk.

The current custom build of Debian is around 300 MB, which includes all the required kernel modules, base libraries, and extra libraries needed to comfortably support the Player system and most anything else needed (this includes the Debian-provided version of the OpenCV computer vision library). That is considerably larger than the initial goal and is mostly the result of some extraneous dependencies on GTK+ libraries within Player, which can be removed when time permits. However, the system does boot in just under 10 s, which is quite good. Boot time can be improved by optimizing the operating system.

3.2.3. Device Abstraction

As mentioned previously, the Player system is a "robot device interface and server" and acts as a device abstraction layer to elements of the robotic architecture residing in the core and brain layers. Since Player is now one of the most popular open source robotics libraries, it already has support for many of the devices used, including IEEE 1394 cameras, USB cameras, the SR3000 flash ladar, and the Microstrain inertial measurement unit (IMU). Adding a device is straightforward—pick or develop a Player interface that defines an abstract representation of a device (e.g., the laser interface defines how to talk to a laser-ranging sensor without worrying about the particulars of device initialization, configuration, or communications protocol). Most drivers then provide a specific implementation of an existing interface and a configuration file format for specifying hardware-specific parameters in a runtime-configurable format.

The ATRV Jr. currently has one custom device driver implementation for the Roboteq motor controller, described in the next section.

3.2.3.1. Roboteq Device Driver

The Roboteq device driver written by UVTD implements the Player position2d interface, which can be used to control planar mobile robots. The interface provides the facility to issue velocity commands (x_dot, y_dot, theta_dot), position commands (x,y,theta), speed/heading commands (v,theta), and car commands (v,theta), which the driver may ignore or implement according to the platform configuration. The current version of the Roboteq driver only understands the velocity commands and assumes the presence of encoders and the use of mixed mode, closed-loop serial operation to the actual motor controller device.

3.2.3.2. Player Configuration

Player is very flexible and does not assume or impose much on a system design. The core of Player is the device abstraction, but Player also provides a Transmission Control Protocol (TCP)-based server that allows multiple remote connections and controls for each device configured for that server. In most cases, there is one Player server per robot providing a connection to all the devices configured for that robot. However, the server is not implemented in a concurrent manner; therefore, the configuration in use on the ATRV Jr. takes advantage of multiple CPUs by providing one Player "server" per robotic device, e.g., a stereo vision server, an IMU server, etc. This provides the same abstraction as one server for all but allows the servers to run concurrently (and therefore block concurrently, if need be).

3.3. Development Environment

Building an essentially new robotic system from the ground up[8] requires that configuration management (CM) and software engineering issues be addressed. Section 3.3.1 highlights the motivation and derived requirements that guide the design of the environment described in section 3.3.2.

3.3.1. Motivation and Requirements

Creating a formal CM environment is motivated by a desire to do the following:

- Keep the robot systems clean and free from version incompatibilities.
- Allow new engineers/developers to become productive with the available tools.
- Facilitate access to the code.
- Share the maintenance and development of the system across the set of contributors.
- Support more than one robot system.
- Encourage and/or enforce compliance with software engineering practices in order to improve code quality.

These goals were used to define the following high-level requirements:

1) The robotic system software must be versioned from a central server.
2) The development model will be a host-target configuration.
3) The development systems must have a common set of libraries and tools, and therefore be imaged from a central server.
4) The development systems must have access to the robot system.
5) The central server and development systems must reside on the same network.
6) The central server must provide reliable data storage.
7) At least two laptops must be available for operating, debugging, and logging data from the robots.
8) It must be easy to update the robots with newly developed software.

The system design that implements these requirements is detailed in section 3.3.2. However, two particularly important ramifications of this CM warrant further description in the following sections.

3.3.1.1. Providing a Clean Slate

Since the ATRV Jr. is a shared resource that will be utilized by multiple researchers often investigating somewhat orthogonal topics, providing a clean operating environment is essential. Extraneous software should be kept to a minimum, with an eye toward the essentials that make the robot run. Not only does this leave more storage capacity, but it speeds up the system and reduces the chance that software might conflict with mission-critical functions. While installing a programmer's text editor and all the development libraries seems harmless, it also seems completely extraneous for a robot running a real mission.[9] In addition, the host-target model helps to ensure that the robot does not get out of sync with the repository. Facilities will be provided to build the relevant portions of the system (i.e., base, core, brain) and then transfer those portions to the robot (which enables a "set it and forget it" behavior).

3.3.1.2. Enabling Software Reuse

The ATRV Jr. upgrade is seen as an opportunity to begin the construction of a development environment and software platform conducive to creating state-of-the-art robotic vehicles based on the x86 architecture. Part of accomplishing this goal is providing for effective software reuse. Thus this approach relies on system connectivity, redundant data storage and automated backups, capable version control and a defined usage policy, and modular software design. Systems need to be connected in order to facilitate source code-level sharing and allow the systems to enforce a check-in policy. Reliable data storage is a requirement to prevent loss of valuable work and knowledge. Version control is a must in order to provide a well-known repository of code and a means to keep it organized and safely shareable. Finally, modular software design (see section 3.2.1) is the real key to reuse; while it is clear that a given robot will have some customized pieces of software that are likely unusable in a different robot, it is equally clear that many algorithms, frameworks, and sensor device drivers can be used across many robots. This is ultimately the purpose in open source software packages like the Player/Stage project and Yet Another Robot Platform (YARP) [14]. A very conscious decision was made in this project to consider the packages available and reuse others' hard work as much as possible (i.e., the use of Player/Stage and the Unified System for Automation and Robot Simulation [USARSim] [15] as well as the ideas, algorithms, and/or code from packages like YARP and the Mobility Open Architecture Simulation and Tools [MOAST] [16] framework). This has, without a doubt, accelerated the ATRV Jr.'s software development significantly.

3.3.2. Environment Design

Based on the requirements stated previously, the environment consists of the following elements: an online Ubuntu GNU/Linux workstation, an offline network, a development server, multiple development laptops, robots, a version control system (VCS), a remote synchronization server, and a host of scripts implementing build and update functionality. A high-level diagram of the topology is shown in figure 9.

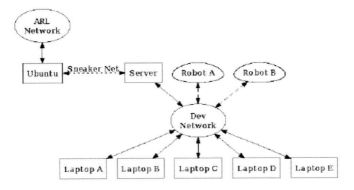

Figure 9. An Ubuntu workstation is online and provides updates to the development server through a manual process, which, in turn, provides automatic updates for the offline network.

3.3.2.1. Online Ubuntu Workstation

Since Ubuntu [17] GNU/Linux is based on Debian GNU/Linux, it inherits the excellent APT and an extensive repository of software.[10] The online Ubuntu workstation provides access to this resource, and custom scripts generate local repository mirrors that can be transferred to the offline development server.

3.3.2.2. Offline Network

An offline network is necessary since the custom robot systems (based on Debian GNU/Linux) and development systems (based on Ubuntu GNU/Linux) cannot currently gain access to the Internet.[11] This network connects the development server and workstations together to provide version control and system software updates, and also allows the workstations to connect to the robots.

3.3.2.3. Development Server

The development server is an Ubuntu GNU/Linux-based system that provides the master VCS repository (on a RAID-1[12] setup providing reliable data storage) as well as all the required libraries and tools for software development. The development server also provides a file synchronization utility (rdist[13]) service for the development laptops and considerable computational power for robot simulations.

3.3.2.4. Development Laptops

The development laptops are intended to be shared resources that serve as both workstations for software development and control stations for system testing. They are configured to enforce a VCS check-in schedule in order to ensure little or no data loss should some hardware fail. The development server automatically provides remote software updates to each laptop; thus all laptops have synchronized file systems. At least two of these laptops can be used specifically as control stations during testing or demonstration.

3.3.2.5. Version Control System

An important part of a basic software development environment is always the VCS. It stores the history of the project and all development artifacts (i.e., it doesn't need to hold only source code). To offer the greatest flexibility, a distributed VCS[14] is utilized. This allows easier branching and merging of individual lines of development (i.e., facilitates experimentation as well as bug fixes for stable releases) and makes it possible to develop outside the network.

Since any "clone" of the repository is actually a full repository in itself,[15] one can transfer the repository to a completely different network or simply take a repository offline for development away from the server. When this "disconnected" repository needs to synchronize with the "connected" repositories, the differences are automatically merged (whether through a network connection or the exchange of patch sets through some medium like e-mail). This configuration allows for the most flexibility, especially for contributors that may not have access to the network.

3.3.2.6. Remote Synchronization System

As mentioned in section 3.3.2.3, the development server hosts a rdist daemon that ensures the file systems across the server and laptops are identical. Note that home directories are not synchronized, while all the important development tools and libraries are kept up to date. This does require some manual updates to be made to the development server itself (see section 3.3.2.2) but only when changes are required (security update or new version of a library, etc.).

3.3.2.7. Build Scripts

A library of build scripts is distributed as part of the robot software distribution's development repository. These build scripts automate and therefore simplify a large portion of the effort needed to construct aspects of the robot software package. For example, there are scripts to download and build the base component as well as scripts to automatically configure the Player library for the ATRV Jr.'s specific device configuration. These scripts are expected to be developed over time to include the most configurable aspects of the system in such a way as to greatly facilitate the construction of a software package for an entirely new piece of robot hardware.

3.3.2.8. On the Host-Target Development Model

The development environment for the new ATRV Jr. is based on a host-target environment, meaning a large portion of development happens on a developer's workstation or a stand-in platform, and the resulting product is transferred to the robot for testing. While this is a departure from the previous ATRV configuration that allowed self-hosted development, it is not as bad as it seems. By using the Player framework for hardware device abstraction, researchers get to make use of supported simulation environments, including (but not necessarily limited to) Stage, Gazebo, and USARSim. A debugger is present in the default base distribution and the host-target separation can be

3.4. Toward a Robotic Agent Architecture

relaxed for situations where it makes more sense to develop directly on the robot (however, review section 3.3.1.1 for reasons to avoid that scenario).

3.4. Toward a Robotic Agent Architecture

While the software support on the ATRV Jr. can, at present, be used for application development and provides a high-level division between modules responsible for different tasks, it does not address how individual components work together to achieve the total behavior for the system. The previous ATRV Jr. software architecture is similar to the current one in that it is based on a collection of services using the Common Object Request Broker Architecture. While services can go a long way in providing reusable device abstractions and behaviors, the architecture concepts discussed in the next sections extend that paradigm to a higher level—that of cooperating agents.

3.4.1. Issues

In the context of developing a framework to control the ATRV Jr. platform, the term architecture refers to how the overall behavior, intelligence, and control of the ATRV will be arranged in software. The term "agent" refers to a self-contained entity that implements a perceive -> think -> act loop and can communicate with other agents. The idea of the agent architecture is to implement the "brain" as a collection of loosely coupled agents acting individually to perform specific tasks and concurrently acting together to enable the emergent behavior, intelligence, and control of the robot. Since an agent can be thought of as a superset of a service, an agent-based architecture can parallelize and distribute across multiple CPUs or systems as well as service architectures can. An agent-based architecture also distributes better than a monolithic system (which cannot run across multiple systems at all).

The concept of agents is not particularly new or unique. Many robot software implementations provide the capability to construct stand-alone components that can communicate with other components, sometimes within a prescribed methodology, sometimes without any guidelines for application structure at all, and almost always tied to a specific implementation language. The following questions are addressed with a new architecture design based on agents:

- What infrastructure and performance capabilities are required to effectively support advanced perception research?

- What infrastructure is required to support autonomous behavior research?
- Is there a way to integrate concepts from the cognitive artificial intelligence perspective into a cohesive software framework that can also include more primitive capabilities?
- Can the architecture be efficient enough to run aboard relatively small robots but flexible enough to support more advanced software designs (i.e., distribution, concurrency, clustering)?
- Is there a way to integrate existing software as a functioning component within the architecture?

3.4.2. Initial Design Topics

The architecture is broken down at the highest level into two components: the infrastructure and the interfaces. There really is a third component, the actual implementation of the agents, but it is orthogonal to the architecture design. Note that hardware abstraction is handled by the Player software and is therefore omitted in the following discussion.

3.4.2.1. Infrastructure

The infrastructure component includes everything that enables the agents to cooperate and perform their functions but includes and prescribes nothing related to actual robot behavior. In other words, it acts as a substrate that supports the agents both during development and runtime.

Initial requirements for the infrastructure include:

- be as lightweight as possible
- enable multilanguage agent implementations
- support appropriate peer-to-peer and service-oriented constructs
  - o discovery and lookup
  - o group communication/multicast
- utilize message-passing abstractions for all inter-agent communication
- allow prioritized messages to disambiguate conflicting requests
- provide efficient implementation constructs where possible
- provide a simple method for launching agents

3.4.2.2. Interfaces

The interfaces describe the kinds of agents or services one uses to build a robotic intelligence system (but not necessarily how to build the agents). By specifying interfaces, one can effectively describe the requirements of an agent without specifying implementation details and decoupling agents from other agent implementations, providing for a more fluid system design that allows for parallelism, distribution, multilanguage support, and pluggable algorithms. For example, table 1 lists some agents and services researchers would like to implement on the ATRV Jr.

3.4.3. Existing Work

The following sections provide brief surveys of a number of existing robot software systems that may serve as foundations or guidelines for future work.

3.4.3.1. Player

Player provides a multiclient/server paradigm over TCP for interacting with robot hardware and does not impose or suggest any other structure or organization. The common use case involves a single server representing the devices on a robot and one or more client programs implementing behavior. Player is written in C/C++ and provides C, C++, and Python client interfaces directly, while others have provided a host of interfaces for other languages (e.g., Java, Octave, Matlab). Refer to section 3.2 for information on the way this project is already leveraging the capabilities of this software package.

Table 1. A representative example of agents and services for the ATRV Jr., roughly organized by complexity

Low	Middle	High
Safety	Identification	Planning
Mobility	Tracking	Task
Mapping	Telemetry/reporting	Health
Navigation	Manipulation	Cooperation
Geometry	Memory	Metareasoning
Obstacle	Context	Human interaction

3.4.3.2. The Mobility Open Architecture Simulation and Tools (MOAST)

MOAST framework is based on the four-dimensional real-time control system architecture developed at the National Institute of Standards and Technology [19]. The framework divides functionality into vertical hierarchies called echelons (e.g., primitive echelon, autonomous mobility echelon), where each echelon is further divided into functional components (e.g., sensor processing, world modeling). The Neutral Messaging Language is used for platform-independent communication between modules (which can be distributed across different systems). Like most of the systems described here, MOAST is open source.

MOAST is one of the most promising candidates for integration into this project if further investigation indicates it directly supports or allows development of the needed infrastructure and interfaces outlined in section 3.4.2.

3.4.3.3. Coupled-Layer Architecture for Robotic Autonomy (CLARAty)

The CLARAty project is described as a reusable robotic software framework. While not truly open source, this project has many algorithm implementations that may be useful as standalone elements in a new system. Integration may be difficult, however, since while the design is very modular, it does not seem to support concurrency or distribution explicitly.

3.4.3.4. Other Packages

While investigating previous work, the authors have come across a large number of software packages aimed at developing mobile robots, including but not limited to YARP, RobotCub, Saphira, OROCOS, MARIE, FlowDesigner, and RobotFlow. Descriptions of these packages are outside the scope of this report; a future report describing the proposed agent architectures will provide more in-depth discussion of the state of the art.

4. CONCLUSION

The ATRV, even with its upgrades, is not by any means a military robot. It is a platform that provides infrastructure (mobility, power, communication) supporting the research elements of the program, i.e., perception processing and autonomous behaviors. As these behaviors take shape and mature, they will be transitioned to robots designed for the field, along with sensors, platforms, and computing elements coming from other efforts.

The upgraded ATRV has computing power and communications at the state-of-today's practice, a suite of sensors with a variety of technologies and complementary strengths, an architecture built on the foundations of the best robotics research institutions in academe, and safety substantially improved over the baseline. Lessons have been learned in the upgrade, which strengthened the skills and understanding of the researchers involved. It is to be expected that the process, as well as the product, of the upgrade effort will enhance the research efforts of ARL's Unmanned Systems Division for years to come.

REFERENCES

[1] VIA Technologies, Inc., Mini-ITX announcement. http://via.com.tw/en/initiatives/spearhead /mini-itx/ (accessed 1 September 2008).

[2] Alfa Network, AWAP608 product page. http://dplanet.biz/alfa.com/product_info.php?cPath=159_33_55&products_id=156 (accessed 3 September 2008).

[3] Unibrain, Fire-i digital camera product page. http://www.unibrain.com/Products/VisionImg /Fire_i_DC.htm (accessed 27 August 2008).

[4] Maxbotix, LV-MaxSonar-WR1 Beam Plots. http://www.maxbotix.com/Performance_Data .html (accessed 27 August 2008).

[5] Acroname Robotics, URG-04LX product page. http://www.acroname.com/robotics/parts /R283-HOKUYO-LASER1.pdf (accessed 27 August 2008).

[6] Mesa Imaging, SR3000 product page. http://www.mesa-imaging.ch/pdf/SR3000_Flyer _Sept07.pdf (accessed 27 August 2008).
Videre Design, STOC product page.http://www.videredesign.com/vision/stoc.htm (accessed 27 August 2008).

[7] MicroStrain, Inc. 3DM-GX1 product page. http://www.microstrain.com/3dm-gx1.aspx (accessed 27 August 2008).
Mp3Car.com, M2-ATX product page.http://store.mp3car.com/M2_ATX_160W_Intelligent _DC_DC_PSU_p/pwr-015.htm (accessed 27 August 2008).

[8] Roboteq, AX3500 product page. http://www.roboteq.com/ax3500-folder.html (accessed 27 August 2008).

[9] Gerkey, B.; Vaughan, R. T.; Howard, A. The Player/Stage Project: Tools for Multi-Robot and Distributed Sensor Systems. *Proceedings of the 11th International Conference on Advanced Robotics*, Coimbra, Portugal, June 2003; pp 317–323.

[10] Player Wikipedia, Player Users Wiki page. http://playerstage.source forge.net/wiki/PlayerUsers (accessed 1 September 2008).

[11] Debian Home Page. http://www.debian.org/ (accessed 3 September 2008).

[12] YARP Home Page. http://eris.liralab.it/yarpdoc/index.html (accessed 1 September 2008).

[13] SourceForge.net, USARSim description. http://sourceforge.net/projects/usarsim (accessed 1 September 2008).

[14] SourceForge.net, MOAST description. http://sourceforge.net/projects/moast/ (accessed 1 September 2008).

[15] Ubuntu Home Page. http://www.ubuntu.com/ (accessed 3 September 2008).

[16] MagniComp, Rdist Home Page. http://www.magnicomp.com/rdist/ (accessed September 2008).

[17] Scrapper, C.; Balakirsky, S.; Messina, E. MOAST and USARSim: A Combined Framework for the Development and Testing of Autonomous Systems. *Proceedings of the SPIE Defense and Security Symposium*, Orlando, FL, April 2006.

APPENDIX. SENSOR SELECTION FOR A FAST INDOOR ROBOT

This appendix describes mobility sensor requirements for a robot capable of speed comparable to that of a human in similar circumstances. The assumed domain is indoors or mild outdoor terrain. The implication of indoor terrain is that there is little room for evasive action (a hallway might end in stairs that span the entire width of the hallway), so the most critical response to a hazard is to stop. Sensors must be able to detect an incipient hazard at a range sufficient to allow room for a complete stop.

The speed at which a human moves indoors can be parameterized as follows. Human walking speed is roughly 2 m/s, while human running speed can be roughly bounded by the rate of a world-class sprinter, say 10 m/s. Indoors, it seems unlikely that a human will move at a sprinter's pace, so assume a maximum speed of 5 m/s.

The deceleration of the robot to a complete stop is governed by brakes (for a vehicle-like robot) and coefficient of friction. For design purposes, the coefficient of friction is assumed to be 0.5, limiting deceleration to roughly 5 m/s/s.

The distance required for the robot to stop can now be calculated. From 5 m/s, with braking acceleration limited by the coefficient of friction, it will take the robot 1 s to decelerate to a standstill, during which time it will cover a 2.5-m distance. The implication for the sensor suite is that the robot must be able to reliably detect obstacles at a distance of at least 2.5 m. A safety margin for reaction/response time should also be included. A human can react in 0.1 s; if a robot reacts in the same time (not necessarily a conservative assumption), an additional 0.5 m must be added to the sensor detection range for a total of 3 m.

In general, sensors have an easier time detecting a positive (above the local surface) than a negative (below the local surface, e.g., a hole or depression) obstacle. However, both kinds of obstacles must be detected.

DISCLAIMERS

The findings in this report are not to be construed as an official Department of the Army position unless so designated by other authorized documents.

Citation of manufacturer's or trade names does not constitute an official endorsement or approval of the use thereof.

End Notes

[1] The products described in this report are believed to be suitable for the intended use, but their use in this endeavor does not constitute an endorsement by the government. Other products may perform as well or better, or be less expensive.

[2] Model No. LV-676.

[3] Usually an 8-bit digital interface.

[4] A computer bus for attaching mass storage devices to a computer.

[5] Also known as Firewire (another serial bus standard for computer interfaces).

[6] A standard serial interface to a computer, once common but recently supplanted by USB.

[7] A standard for computer display hardware.

[8] At least from the software point of view.

[9] Recently, there has been some discussion on relaxing these restrictions through careful build-time parameters; e.g., a build switch could indicate whether the robot is being used for development or "production" use.

[10] Often, software is not available anywhere else without a manual installation procedure: find dependencies, download, compile, and install everything in the proper order (a time-consuming process).

[11] For reasons beyond the scope of this report.

[12] Raid = redundant array of independent disks.

[13] Remote file distribution: a method of distributing software updates from a central location to multiple remote sites. The behavior is implemented with a client (rdist) and a server (rdistd) (*18*).

[14] Currently, Mercurial is the distributed VCS of choice. However, Git is also being investigated for its power and flexibility.

[15] Contrast this with centralized, nondistributed VCSs like CVS and Subversion (one must have access to the server in order to check in code).

In: Robotic Autonomy and Control
Editors: T.C. Mueller and M.E. Bynes

ISBN: 978-1-62100-605-3
©2012 Nova Science Publishers, Inc.

Chapter 2

SCALABILITY OF ROBOTIC CONTROLLERS: EFFECTS OF PROGRESSIVE LEVELS OF AUTONOMY ON ROBOTIC RECONNAISSANCE TASKS[*]

Rodger A. Pettitt, Elizabeth S. Redden, Estrellina Pacis and Christian B. Carstens

1. INTRODUCTION

1.1. Background

Maes (1995) proposed a continuum of robot control that ranges from teleoperation to full autonomy. Robots can be placed on this continuum based on the level of human-robot interaction that is required. Teleoperation is the lowest level of automation on the continuum because it requires the most intervention from the robot operator. A teleoperated robot is totally under the control of the operator who uses a joystick or other control device to command the robot. This requires constant interaction between the robot and the operator. Teleoperation is not the ideal solution for all situations. Many

---

[*] This is an edited, reformatted and augmented version of an Army Research Laboratory publication, ARL-TR-5258, dated August 2010.

teleoperation tasks are repetitive and boring and the work requires constant attention by the operator.

Semi-autonomous control (often called supervisory control) requires the operator to provide an instruction or portion of a task that can safely be performed by the robot on its own. Two types of semi-autonomous control are often identified—shared control (also called continuous assistance) and control trading. Shared control requires the teleoperator to delegate a task for the robot or to accomplish it via direct control of the robot. If the operator delegates control to the robot, he or she must still monitor the robot to ensure that it is performing the task correctly. A guarded teleoperated robot can be placed on this portion of the continuum because it has the ability to sense and avoid obstacles but will otherwise navigate as driven, like a robot under manual teleoperation. During control trading, the human only interacts with the robot to give it a new command or to interrupt it and change its orders. A line-following robot can be placed on this portion of the continuum if it simply follows something painted, embedded or placed on the floor and does not have the ability to circumnavigate obstacles on its own.

After receiving instructions, autonomous robots (those that perform tasks independently and/or have the capacity to choose goals for themselves) operate under all reasonable conditions without recourse to an outside operator and can handle unpredictable events (Haselager, 2005). Huang, et al. (2005) state that a fully autonomous robot has the ability to gain information about the environment, work for an extended period without human intervention, move itself through its environment without human assistance, and avoid situations that are harmful to people, property, or itself unless those are part of its design specifications. An autonomously guided robot can be placed on this portion of the continuum because it knows at least some information about where it is and how to reach waypoints (Murphy, 2000). Knowledge of its current location is determined by using sensors such as lasers and global positioning systems (GPS). Positioning systems determine the location and orientation of the platform, so the robots can plan a path to their next waypoints or goals. Autonomous robots make their own choices instead of following goals set by other agents. A robot that is capable of goal generation can also be placed on this portion of the continuum.

The incorporation of autonomous robotic systems into military schema involves more than just effective system design. It is also important for designers to understand which potential robotic system autonomous capabilities match or surpass human abilities and when, during scenario accomplishment, the human needs assistance or is overloaded. This knowledge is needed in order to leverage autonomous robotic behaviors for optimal performance. Thus the goal is to optimize the human and robot roles in a task. To do this, robot control competencies and inefficiencies must first be identified and then they must be understood in relation to specific task performance. Trade-offs among levels of autonomy must be identified. For example, an autonomous system may have a slower reaction time to difficult problems than a system being teleoperated by a human but latency involved in teleoperation could render the human's reaction time ineffective. Also, moderators of robot control performance must be identified. For example, workload is also an important moderator of human task performance. The effectiveness of a control system is likely to depend a great deal on situational task demands. The requirement for the operator to perform simultaneous tasks such as controlling multiple robots or performing local security tasks could have a detrimental effect on human intervention in robotic task outcomes because of human availability at a certain point in time and because of cognitive overload.

Many studies have demonstrated that operators' situation awareness (SA) was higher when they were controlling robots with semiautonomous or autonomous capabilities (Chen et al., 2008; Dixon et al., 2003; Luck et al., 2006). However, increased autonomy is not a panacea. In a two-year study of a collaborative human-robot system Stubbs et al. (2007) found that as autonomy increased, users' inability to understand the reasons for the robot's actions disrupted the creation of common ground. Also, many fear that providing more and more autonomy to armed robots could result in collateral damage or fratricide (Singer, 2009).

1.2. Objective

The goal of this research was to examine the effectiveness of different levels of automation (teleoperation, semi-autonomous, and autonomous) on robot control in reconnaissance missions when the robotic operator is fully engaged in additional high cognitive load activities. Our hypothesis was that full autonomy would be the most effective level of autonomy for a robotic reconnaissance mission when the robotic operator is fully engaged in a high cognitive load activity.

1.3. Overview of the Experiment

This study was a cooperative research effort between the U.S. Army Research Laboratory (ARL)/Human Research and Engineering Directorate (HRED) and the Space and Naval Warfare Systems Center San Diego (SSC Pacific). It was an investigation of the effect of progressive levels of automation on robotic reconnaissance task performance. It took place at Fort Benning, GA. Thirty Soldiers from the Officer Candidate School (OCS) and the Warrior Training Center (WTC) participated in the study. After training on the operation of the robotic system, each Soldier completed reconnaissance exercises using three different levels of robotic automation (teleoperated, semi-autonomous and autonomous). During the exercises, Soldiers responded to requests for information regarding situation and mission awareness. The terrain and hazards were counter-balanced along with the automation level to control for the effect of learning. Automation level and usability were evaluated based on objective performance data, data collector observations, and Soldier questionnaires.

2. METHOD

2.1. Participants

Thirty Soldiers from the OCS and WTC participated in the study. The OCS participants included Soldiers with prior enlisted service with a variety of backgrounds and experience levels as well as those just coming into the Army from college. The WTC Soldiers consisted of enlisted Soldiers with the rank of E-4 through E-6 serving as instructors for the pre-Ranger and air assault courses.

2.1.1. Pretest Orientation

The Soldiers were given an orientation on the purpose of the study and what their participation would involve. They were briefed on the objectives and procedures, as well as on the robot. They were also told how the results would be used and the benefits the military could expect from this investigation. Any questions the subjects had regarding the study were answered.

2.2 Apparatus and Instruments

2.2.1 SSC Pacific Robot

The SSC Pacific robot (figure 1) is an iRobot PackBot Scout, equipped with a 1st generation Navigator payload. The payload contains a sensor suite that includes an inertial measurement unit (IMU), a gyroscope, a compass, GPS, and a 360° ladar. The IMU, gyroscope, compass, and GPS are used for positioning, localization, and navigation, while the ladar is used for obstacle avoidance and mapping. In addition, the payload contains a processor that runs the Autonomous Capability Suite (ACS) that includes all the autonomous behaviors onboard the robot and Freewave radio to communicate with the operator control unit. The scalability and configurable framework of ACS provided the key sliding autonomy feature needed to conduct this experiment; the operator was able to change the level of autonomy onboard the robot by a press of a button on the operator interface.

Figure 1. SSC Pacific robot with 1st generation navigator payload.

2.2.2. Operator Interface

The operator interface used to control the system was based on SSC Pacific's Multi-Robot Operator Control Unit (MOCU). An example screenshot of the interface is found in figure 2. The robot's location, driven path, goal points, and sensor data (i.e., map data) were overlaid on an aerial image. Real-time video from the robot was also displayed.

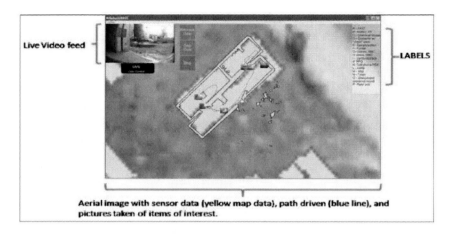

Aerial image with sensor data (yellow map data), path driven (blue line), and pictures taken of items of interest.

Figure 2. Snapshot of MOCU interface.

Various button and joystick controls were also provided to send drive commands to the robot, as well as to turn on and off the robot's behaviors. This allowed the operator to control the sliding autonomy available on the robot. Three levels of autonomy were achieved by turning on and off robot behaviors. All behaviors were turned off during teleoperation so that the operator was in complete control of all robot actions and the operator had to draw the building map and mark the location of items of interest by hand. Semi-autonomy was achieved by using the obstacle avoidance, mapping, and return-to-start functions available on the robot. This function assisted the robot operator during driving by automatically avoiding obstacles, self directing itself to open spaces, building a map of the building for the operator, and returning to the start location on its own when the operator completed his/her mission. Full autonomy was achieved by turning on the efficient exploration behavior in addition to all the behaviors included under semi-autonomy. Table 1 shows the behaviors used for each level of autonomy.

Table 1. Autonomous behaviors used for each level of autonomy

	Self-Directing	Obstacle Avoidance	Mapping	Return-to-Start	Exploration
Teleoperated	No	No	No	No	No
Semi-autonomous	Yes	Yes	Yes	Yes	No
Fully-autonomous	Yes	Yes	Yes	Yes	Yes

The MOCU software was developed independent of the hardware unit, and thus can be run on any computer. A commercial grade off-the-shelf (COTS) laptop was used in the experiment to run MOCU. A Microsoft Xbox 360 wireless (figure 3) controller was used to provide a joystick/button interface.

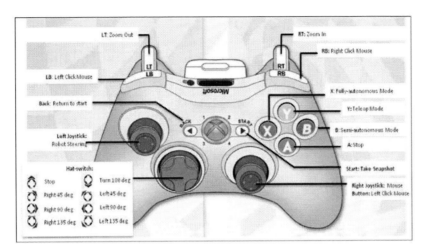

Figure 3. Microsoft Xbox 360 wireless controller.

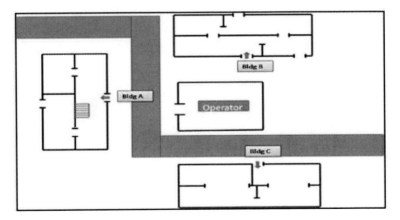

Figure 4. Building reconnaissance course.

2.2.3. Building Reconnaissance Course

The building reconnaissance course (figure 4) was located at Molnar military operations in urban terrain (MOUT) site, Ft Benning, GA. It consisted of three one-story buildings that were similar in size but with different floor plans. Soldier operators were located out of the line of sight of the robot in a stationary position inside a separate building.

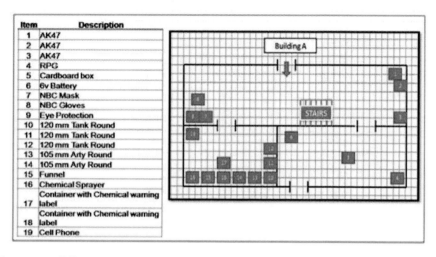

Figure 5. Building A: chemical weapons factory.

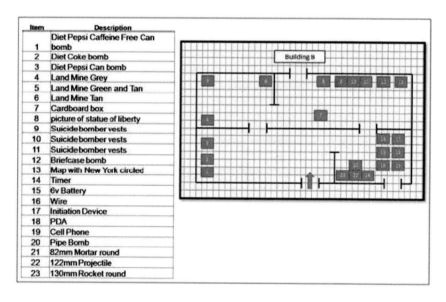

Figure 6. Building B: terrorist staging area.

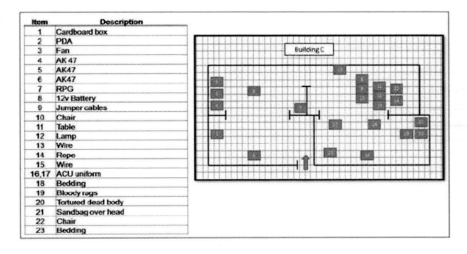

Figure 7. Building C: POW interrogation facility.

Various items were placed in the rooms of each building and arranged to represent a specific enemy activity. Building A was configured to represent a chemical weapons factory, building B was configured to represent a staging area for a terrorist attack against the United States, and building C was configured to represent a prisoner of war (POW) interrogation facility. Figures 5-7 show the floor plans and placement of items for each building.

2.2.4. The National Aeronautics and Space Administration-Task Load Index (NASA-TLX)

The NASA-TLX is a subjective workload assessment tool that allows subjective workload assessments on operators working with various human-machine systems (Hart and Staveland, 1988). It uses a multi-dimensional rating procedure that derives an overall workload score based on a weighted average of ratings on six subscales. These subscales include mental demands, physical demands, temporal demands, own performance, effort, and frustration. It can be used to assess workload in various human-machine environments such as aircraft cockpits; command, control, and communication workstations; supervisory and process control environments; simulations; and laboratory tests. The version of the NASA-TLX used during this experiment was presented to the Soldiers on a computer. Definitions of each scale were provided on laminated paper so the participates could refer to it as they were providing their estimates of the workload associated with each type of small unmanned ground vehicle (SUGV) control on the various scales.

2.2.5. Questionnaires

A demographic questionnaire was administered to gather information concerning demographic data, robotic experience, and physical characteristics that might have had an effect on the participant's ability to operate the robot. The questionnaires given after participation in the events were designed to elicit Soldiers' opinions about their performance and experiences with each of the control systems. They asked the Soldiers to rate the devices on a 7-point semantic differential scale ranging from "extremely good/easy" to "extremely bad/difficult." A post iteration questionnaire was administered to each Soldier at the end of each iteration with each level of autonomy, and an end of experiment questionnaire was administered at the completion of all three iterations.

2.3. Procedures

2.3.1. Soldier Orientation

Upon arrival, the experiment Soldiers received a roster number, which was used to identify them throughout the evaluation. During the pretest orientation, the Soldiers were given an operations order that explained the robotic mission that they would be undertaking during the experiment. A copy of the operations order can be found in appendix A (FM 5-170, 1998). It was explained to them that upon completion of training they would complete three iterations of the building reconnaissance course (one with each level of autonomy) according to the matrix in table 2. All instruments that were used, the training course, and the building reconnaissance course were explained and any questions the Soldiers had concerning the experiment were answered. They were also told how the results would be used and the benefits the military could expect from the investigation. Demographic data, as well as data concerning their Army and robotic experience and physical characteristics, were also taken for each Soldier.

2.3.2. Demographics

Demographic data were taken for each Soldier. Data concerning their physical characteristics and experience, especially their knowledge of operating remote controlled vehicles, were included in the demographic data sheet. The demographic data sheet is shown in appendix B.

2.3.3. Training

Training was conducted by a representative of SSC Pacific. Groups of three Soldiers were given an overview of the robot capabilities, controls, and sensors, and the three autonomy levels were explained. Then each Soldier was given 15 min of individual hands-on training that included practice on the tasks required to perform the building reconnaissance in all three levels of autonomy. The trainer assessed whether the Soldiers were capable of performing the tasks in all three levels of autonomy before they were allowed to participate in the study.

During training, Soldiers were issued a battalion level operations order outlining the reconnaissance mission they were supporting. Each Soldier was given about 15 min to read the order prior to conducting the reconnaissance course.

Table 2. Order of treatments and lanes

Roster Number	Building and Iteration		
	Bldg A, Trial 1	Bldg B, Trial 2	Bldg C, Trial 3
1, 16	T	S	A
2,17	A	T	S
3,18	S	A	T
4,19	T	A	S
5,20	S	T	A
6,21	A	S	T
7,22	S	T	A
8,23	A	S	T
9,24	T	A	S
10,25	S	A	T
11,26	T	S	A
12,27	A	T	S
13,28	S	A	T
14,29	A	T	S
15,30	T	S	A

Note: T = teleoperation; S = semi-autonomous; A = autonomous.

2.3.4. Building Reconnaissance Course

The Soldiers completed the building reconnaissance course with the level of autonomy and the lane order assigned in table 2. The course consisted of three different one-story buildings so the Soldiers could reconnoiter a different building with each level of autonomy.

During teleoperation trials, the robot was manually driven using the video feed displayed on the MOCU interface and left joystick gimbal on the Microsoft Xbox 360 wireless controller (figure 3).

During the semi-autonomous trials, the robot's obstacle avoidance, mapping, and return-to-start behaviors were turned on, and the robot self-directed itself through the building to the next open space available. The Soldier was able to stop the robot from self-directing itself and take control at any time by pressing the A button to stop the robot and used the left joystick gimbal to influence the robot's direction. To turn on the robot's self-directing behavior again, the Soldier pressed the B button. The obstacle avoidance feature was always on to prevent the robot from running into objects. To turn obstacle avoidance off, the Soldier had to switch to teleoperation mode by pressing the Y button. When the Soldier decided all items of interest were found, the BACK button was pressed on the X-box controller and the robot returned to the starting location on its own.

During the fully-autonomous trials, the robot's obstacle avoidance, mapping, and return-to-start behaviors were turned on, as well as an efficient exploration behavior that helped the robot self direct itself in places it had not driven to yet, to minimize the overall time to explore the building while maximizing coverage area. As in the semi-autonomous mode, the Soldier was able to stop the robot at any time by pressing the A button. The Soldier could also switch teleoperation mode (by pressing the A button) or semi-autonomous mode (by pressing the X button). Also as in the semi-autonomous mode, the Soldier was able to send the robot back to the starting location by pressing the BACK button.

For all three trials in each of the autonomy levels, the Soldier was required to take snapshots of items of interest found and label them. The camera was in a fixed position facing the front of the robot and did not have a pan and tilt capability. In order to take a snapshot, the Soldier had to maneuver the robot into a position facing the item. To label the snapshots, the Soldier used the right joystick gimbal to move the mouse to the snapshot, click the right trigger button on the X-box controller to bring up the list of labels, and then click the left trigger button to select the label.

Soldiers were instructed that their primary tasks were to provide a map of the floor plan of the building, to identify objects of interest (SA task) and take a snapshot of them, and return the robot to the starting point as quickly as possible once the entire building had been reconnoitered and mapped. During autonomous and semi-autonomous trials, the system automatically mapped the buildings as the robot maneuvered through them and snapshots of the items appeared on the map in the location they were taken. During teleoperated trials, Soldiers were instructed to draw a scaled map by hand of the floor plan of each building depicting the location of items of interest.

The Soldiers' secondary tasks were to answer questions concerning details of their mission based on the information provided in the operations order, and to identify different types of smoke grenades. The secondary tasks served as an additional cognitive load. Secondary task questions were asked at 2 min intervals throughout the reconnaissance. Mission details could be found in excerpts of the operations order that was provided to them during each trial. They were instructed to look up and answer questions about the operations order only if doing so did not interfere with performance of the primary tasks.

A data collector accompanied the robot to record completion times and driving errors (i.e., running into walls or objects with the robot). Another data collector was present at the operator station to record the number and types of items of interest identified (SA task), and to ensure the Soldiers prioritized their focus on the primary task. An additional data collector asked the Soldiers secondary task questions at two minute intervals and recorded their responses.

Upon the completion of the iteration, the Soldiers were provided with a list of the items of interest they identified and were asked to answer situational awareness questions concerning the type of enemy activity that had taken place in the building based on the items found in the building. They were also asked to fill out questionnaires concerning the trial they just completed and to fill out a NASA-TLX concerning the level of workload experienced.

2.3.5. End of Iteration Questionnaire Administration

Questionnaires, designed to assess participants' performance and experiences with each level of autonomy, and the NASA-TLX were administered to each Soldier at the end of each iteration. After completing the course with each level of autonomy, the Soldiers completed an end of experiment questionnaire that compared each condition.

2.4. Experimental Design

The design of this experiment was a single factor repeated measures design (table 2).

2.4.1. Independent Variable
- Level of autonomy (reconnaissance course events).

2.4.2. Dependent Variables
- Building reconnaissance course completion time.
- The number of driving errors on each course.
- The number of correct objects of interest found on each course.
- Accuracy of maps created during the building reconnaissance course (correct number of doors, items of interest in correct relation to each other, correct number of rooms).
- SA questionnaire responses.
- Number of secondary task questions answered correctly.
- NASA-TLX scores for each level of automation after each course is completed.
- Data collector comments.
- Questionnaire ratings and comments.

2.5. Data Analysis

All objective data collected on the reconnaissance courses were analyzed using repeated measures analysis of variance (ANOVA). Follow-on pair wise comparisons were done using Holm's Bonferroni procedure to control for family-wise error rates (Holm, 1979). Partial eta squared (η^2_p), an index of effect size, was computed for each ANOVA. Iteration effects were controlled through the counterbalanced order of the experimental design (table 2). Soldier questionnaire data were analyzed using descriptive statistics on the subjective ratings.

3. RESULTS

3.1. Demographics

The Soldiers ranged in rank from E4 to E6. The average age of the Soldiers was 28 years and the average time in service was 68 months. None of the Soldiers had any prior experience in teleoperating a ground unmanned robot. Detailed responses to the demographics questionnaire are available in appendix B.

3.2. Training

The participants rated the training as being very good for all levels of autonomy. They indicated that the hardest task to learn was mapping the building in the teleoperation mode. Driving the robot was the easiest task to learn in all three levels of autonomy. Detailed responses to the training questionnaire are available in appendix C.

3.3. Building Reconnaissance Course Results

3.3.1. Robotic Control Results

Table 3 and figure 8 show the mean times to complete the course using each of the operation modalities. A repeated measures ANOVA showed that there was a significant difference among the means, $F(2,58) = 17.4$, $p<0.001$, $\eta 2 p = 0.383$. Follow-on paired-sample t-test comparisons were conducted using Holm's sequential Bonferroni correction for family-wise error rate. As shown in table 4, the average course completion time in the teleoperation condition was significantly slower than average times for the semi-autonomous and autonomous conditions. The slower times demonstrated during the teleoperation trials can be attributed to the requirement to manually map the buildings as well as an increased driving error rate. The average course completion times in the semi-autonomous and autonomous conditions approached significance ($p = 0.069$) with the semi-autonomous condition being slower than the autonomous condition.

Table 3. Mean course completion times (min:sec)

Condition	Mean	SD
Teleoperation	16:01	3:54
Semi-autonomous	12:40	4:06
Autonomous	11:21	3:09

Table 4. Follow-on paired comparisons, course completion times

Pair	t	df	Obtained p	Required p
Tele vs. Semi	3.77	28	0.001a	0.025
Tele vs. Auto	5.75	29	<0.001a	0.0167
Semi vs. Auto	1.89	28	0.069	0.05

[a] $p<0.05$, 2-tailed.

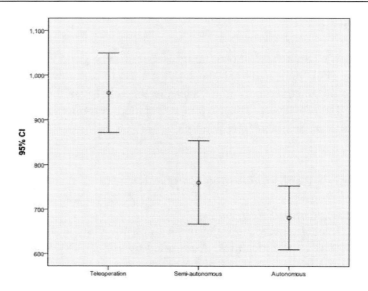

Figure 8. Mean course completion times (s) with 95% confidence intervals.

Table 5 and figure 9 show the mean number of driving errors with the three operation modalities. A repeated measures ANOVA indicated that there was a significant difference among the means, $F(2,58) = 7.24$, $p = 0.002$, $\eta^2_p = 0.200$. Follow-on paired comparisons, summarized below in table 6, indicated that there were significantly more driving errors in the teleoperation condition than in the semi-autonomous and autonomous control conditions.

Table 5. Mean driving errors

Condition	Mean	SD
Teleoperation	3.67	3.33
Semi-autonomous	1.33	2.25
Autonomous	1.90	2.32

Table 6. Follow-on paired comparisons, driving errors

Pair	t	df	obtained p	required p
Tele vs. Semi	4.14	29	<0.001a	0.0167
Tele vs. Auto	2.52	29	0.017a	0.025
Semi vs. Auto	−0.88	29	0.388	0.05

[a] $p<0.05$, 2-tailed.

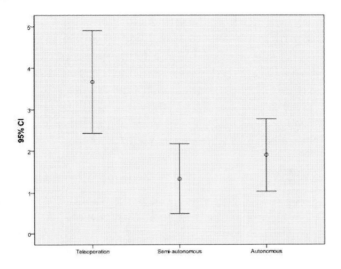

Figure 9. Mean driving errors with 95% confidence intervals.

Based on the post iteration questionnaire results, Soldiers rated the task of mapping the building during teleoperation trials as the most difficult task of the reconnaissance. In the autonomous and semi-autonomous conditions, an accurate map of the building was drawn on the display as the robot moved through the building. In the teleoperation condition the participants were given lined graph paper and asked to draw a diagram of the building, complete with doors and items found in the building. Because the quality of some of the drawings was so poor, it was difficult to interpret the drawings. We conducted a reliability check by having two independent raters code a random sample of fifteen diagrams for the number of doors, the number of doors in the correct location, and the total area (ft^2) of the building. The reliability coefficients are shown in table 7. Although all three correlation coefficients were statistically significant, there was substantial disagreement between the two raters in regard to the number and location of doors in the diagrams.

Table 7. Inter-rater reliabilities, teleoperation maps

Measure	r	p
No. doors	0.65	0.009
No. doors correct wall	0.74	0.002
Area	0.92	0.000

The summary statistics from the map drawing ratings are shown in table 8. Although a few participants correctly identified all of the doors in the correct location, on average only about 2/3 of the doors were correctly identified and drawn in the correct location. All of the participants underestimated the area of the building, most by a substantial amount. One-sample t-tests were used to compare each mean against a value of 100% that would be obtained using the maps automatically generated in the semi-autonomous and autonomous driving conditions. Each of the means was significantly lower than 100%.

Table 8. Summary statistics, teleoperation maps

Measure	Mean	SD	t	df	p
% doors identified	70%	25%	4.75	14	<0.001
% correct door location	67%	27%	4.70	14	<0.001
Area	40%	20%	11.83	14	<0.001

3.3.2. Situation Awareness Results

Table 9 shows the proportion of items detected (perception) in the three buildings for each robotic control condition. A repeated measure ANOVA indicated that there was no significant difference among the means, $F<1.00$.

Table 9. Mean proportion of items detected

Condition	Mean	SD
Teleoperation	74%	14%
Semi-autonomous	75%	16%
Autonomous	72%	11%

3.3.3. Secondary Task Results

During the building reconnaissance trials, the participants were periodically asked questions regarding details contained in the operation order, or they were asked to identify specific types of grenades. Soldiers stated that performing the secondary task during the teleoperation trials was difficult because they had to concentrate more of their attention on driving and manual mapping tasks. Table 10 and figure 10 show the proportion of correct responses to these secondary task questions. A repeated measures ANOVA showed that there was a significant difference among the means, $F(2,58) = 3.79$, $p = 0.028$, $\eta_{2p} = 0.115$. Follow-on paired comparisons, shown in table 11, indicate that secondary task performance was significantly better in the autonomous condition relative to the teleoperation condition.

Table 10. Mean % correct secondary activity

Condition	Mean	SD
Teleoperation	41%	17%
Semi-autonomous	49%	25%
Autonomous	55%	25%

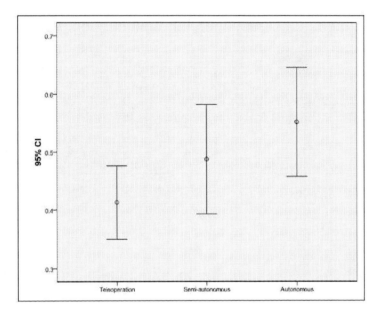

Figure 10. Mean % correct secondary task with 95% confidence intervals.

Table 11. Follow-on paired comparisons, secondary task

Pair	t	df	Obtained p	Required p
Tele vs. Semi	-1.51	29	0.143	0.025
Tele vs. Auto	-2.78	29	0.009[a]	0.0167
Semi vs. Auto	-1.23	29	0.227	0.05

[a] $p < 0.05$, 2-tailed.

3.4. NASA-TLX Results

Table 12 shows the weighted means on the NASA-TLX scales and the total workload means. The means for the three conditions are illustrated in figure 11. Repeated measures ANOVAs on the scales and total workload are summarized in table 13. Follow-on paired comparisons are shown in table 14. There were no significant differences across conditions on the Physical and Performance scales. On all the other scales, and on total workload, the teleoperation condition tended to have the highest workload means and the autonomous condition had the lowest means.

Table 12. NASA-TLX means

Condition	Mental Mean	SD	Physical Mean	SD	Temporal Mean	SD	Performance Mean	SD
Teleoperation	14.13	8.62	1.01	2.38	10.94	7.99	8.58	4.88
Semi-autonomous	12.54	8.75	0.76	2.26	8.54	6.25	8.00	4.91
Autonomous	10.10	8.72	0.87	2.04	6.54	4.71	8.07	5.45

Condition	Effort Mean	SD	Frustration Mean	SD	Total Workload Mean	SD	—	—
Teleoperation	12.58	8.02	9.30	10.79	55.37	22.39	—	—
Semi-autonomous	8.49	7.14	7.46	9.18	45.79	19.70	—	—
Autonomous	7.29	5.97	3.93	6.13	37.82	16.81	—	—

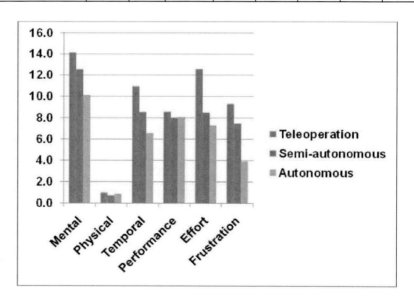

Figure 11. Mean for the three conditions on NASA-TLX.

Table 13. ANOVA summary, NASA-TLX means

Scale	F	df	p	η 2 p
Mental	3.44	2,58	0.039a	0.106
Physical	<1	2,58	0.829	0.006
Temporal	4.32	2,58	0.018a	0.130
Performance	<1	2,58	0.858	0.005
Effort	9.14	2,58	<0.001 a	0.24
Frustration	4.31	2,58	0.018a	0.129
Total workload	9.01	2,56	<0.001a	0.243

[a] $p < 0.05$.

Table 14. Follow-on paired comparisons, NASA-TLX

Scale	Pair	t	df	Obtained p	Required p
Mental	Tele vs. Semi	0.93	29	0.361	0.05
	Tele vs. Auto	2.28	29	0.030	0.025
	Semi vs. Auto	2.30	29	0.029	0.0167
Temporal	Tele vs. Semi	1.47	29	0.152	0.05
	Tele vs. Auto	2.85	29	0.008a	0.0167
	Semi vs. Auto	1.54	29	0.135	0.025
Effort	Tele vs. Semi	3.28	29	0.003a	0.025
	Tele vs. Auto	3.82	29	0.001a	0.0167
	Semi vs. Auto	0.95	29	0.348	0.05
Frustration	Tele vs. Semi	0.98	29	0.337	0.05
	Tele vs. Auto	2.64	29	0.013a	0.0167
	Semi vs. Auto	2.17	29	0.038	0.025
Total Workload	Tele vs. Semi	2.10	29	0.045a	0.05
	Tele vs. Auto	3.67	28	0.001a	0.0167
	Semi vs. Auto	2.80	28	0.009a	0.025

[a] $p < 0.05$, 2-tailed.

3.5. Questionnaire Results

Soldiers were asked to indicate their first, second, and third choice of the level of autonomy they preferred for controlling the robot while performing reconnaissance tasks. Table 15 shows the results of their ratings; the higher the mean, the greater the preference.

Table 15. Level of autonomy preferences for performing reconnaissance tasks

	Mean Rating		
	Teleoperation	Semiautonomous Operation	Autonomous Operation
Route planning	1.62	2.04	2.35
Generating goals	1.88	1.96	2.20
Driving a straight route	1.96	2.15	1.85
Avoiding obstacles	1.81	2.04	2.15
Looking for objects of interest	2.38	1.96	1.81
Overall building reconnaissance	1.73	1.96	2.31

For the most part, the Soldiers' choices for the level of autonomy for reconnaissance tasks indicated that the more autonomy the better. Their choices for such tasks as overall building reconnaissance, avoiding obstacles, generating goals, and route planning showed their preferences for more autonomy. For driving a straight route, the Soldiers preferred semiautonomous operations because they felt that they could drive the robot faster than it traversed in the autonomous mode. They preferred less autonomy when looking for objects of interest because they could spend the amount of time they needed and go closer to the object in order to identify it.

The Soldiers were very positive about the control system because they found it to be easy to learn and to use. Many found it to be familiar because they had spent a lot of time using Xboxes. There were several complaints that the camera on the robot did not have adequate resolution and that they needed to be able to pan, tilt, and zoom the camera in order to identify objects and take pictures more easily. They also experienced some difficulty seeing the display because the room in which they were operating the robot was very bright and glare was a problem. A few Soldiers thought that a color display would assist them in identifying objects. Soldiers also complained about the latency in the system which especially impacted teleoperation.

When asked to choose if they preferred the autonomous or semi-autonomous modes, the Soldiers were closely split, with 16 preferring the autonomous and 14 preferring the semi-autonomous. Those preferring the autonomous mode did so because it freed them to do other tasks and it seemed to work smoother. Those that preferred the semi-autonomous mode did so because they could make the robot move faster and they were able to provide input. Some Soldiers did not like turning over control to the robot at all. Two of them wanted total control at all times. One Soldier commented that the use of both the semi-autonomous and autonomous modes allowed him to shift his focus to the secondary task while maintaining progress toward his primary objective. Due to the small size of the area being reconnoitered and the number of objects in each room, Soldiers constantly had to take manual control and teleoperate the robot into position to take snapshots of the items during the autonomous and semi-autonomous trials.

Drawing an accurate map in the teleoperation condition was especially difficult. The Soldiers were extremely positive about the ability of the robot to build maps of the room and used the map frequently. Soldiers found the mapping capability in the autonomous and semi-autonomous modes particularly useful for gauging their progress while maneuvering through the building.

4. DISCUSSION AND RECOMMENDATIONS

Total vehicle reconnaissance times and driving errors were significantly better when the robot was in the autonomous and semi autonomous modes than when it was in the teleoperation mode. Scribner and Dahn (2008) found very little difference in driving times between teleoperation and their version of semi-autonomy called the biasing mode (the robot semi-autonomously follows a previously recorded GPS path with the operator providing course corrections for GPS path error and obstacle detection and avoidance). However, differences in the levels of autonomy of the robots in the two experiments can explain the different results. The biasing condition was different than the semi-autonomous mode of this experiment because the semi-autonomous mode included autonomous obstacle avoidance and the biasing mode required the operator to perform obstacle avoidance. The biasing mode also did not have nearly the autonomy found in the autonomous mode of this experiment. The latency (the lag between a control input and the response of the robot) complained about by the operators may also have had an impact on the results as it could have had more of a negative impact on teleoperation times and driving errors than on operation in the semi-autonomous and autonomous modes. Lane et al., (2002) indicated that system latency can actually change robotic operator control strategy from continuous command to "move and wait" which adversely impacts reconnaissance times. Over-actuation which is common when latency is unpredictable (Malcolm and Lim, 2003) can also create driving errors.

The consensus of the Soldiers was that mapping the room, driving the robot, and answering the questions in the teleoperation mode created task demands that were too difficult. On the NASATLX, the Soldiers rated teleoperation as creating a higher cognitive and overall work load, higher stress (frustration), more effort, and higher time pressure. While Dixon and Wickens (2003) suggested that automation would relieve cognitive overload, offloading some of the tasks by use of the autonomous modes relieved more than just the cognitive load and made the performance of the tasks much more manageable. This data is supported by Schipani (2003) who stated that workload is increased for higher levels of operator involvement. Map accuracy in the teleoperation mode was much poorer than in the autonomous and semi-autonomous modes. It was clear that the operators' mental models of the environment, based upon viewing it through a robotic driving camera, were fairly inaccurate. This is supported by the findings of Fong et al. (2003) who indicated that operators using teleoperation have difficulty building mental models of remote environments.

Results in the literature concerning SA and targets (objects) identified while the robot is moving are mixed concerning whether higher levels of autonomy result in better SA or vice versa. Chen and Joyner (2006) found that participants detected fewer targets when their robot was operating in the semi-autonomous mode rather than the teleoperation mode. However, automation seemed to benefit UAV pilots' target detection performance in the Dixon et al., (2003) study. Our results were different from both of the previous studies in that no significant differences were found between the levels of automation. However, if SA is more broadly defined in the Chen and Joyner (2006) study to include the gunner task and the communication task, participants did not have better SA in the teleoperation condition. The complex nature of SA and its definition makes it important to specify the definition of SA when making comparisons between studies.

5. CONCLUSIONS

Findings from this experiment indicate that as levels of autonomy increase, workload, reconnaissance times, and driving errors decrease and accuracy of mental maps increase. No differences were found in SA and target identification between levels of autonomy. Future research should be conducted to further define the areas in which automation can improve performance and decrease workload.

DISCLAIMERS

The findings in this report are not to be construed as an official Department of the Army position unless so designated by other authorized documents.

Citation of manufacturer's or trade names does not constitute an official endorsement or approval of the use thereof.

ACKNOWLEDGMENTS

The authors would like to thank the Office of Secretary of Defense Joint Ground Robotics Program under the Urban Environment Exploration Project for their support of this research. We are also grateful for the advice and support of Bart Everett of Unmanned Systems, CAPT

Michael Carter, Gaurav Ahuja, Donnie Fellars, Greg Kogut, and Brandon Sights of SPAWAR Systems Center Pacific, and COL Terry Griffin (Ret), and Joyner Livingston of BFA Systems, Inc. We would also like to express our appreciation to the Soldiers of the Ft. Benning Officer Candidate School and the Soldiers assigned to the Warrior Training Center for their participation in the study. Their voluntary participation and constructive feedback was outstanding and was key in identifying improvements in future robotic platforms.

REFERENCES

Chen, Y. C.; Durlach, P. J.; Sloan, J. A.; Bowens, L. D. Human–Robot Interaction in the Context of Simulated Route Reconnaissance Missions. *Military Psychology* 2008, *20*, 135–149.

Chen, Y. C.; Joyner, C. T. *Concurrent Performance of Gunner's and Robotic Operator's Tasks in a Simulated Mounted Combat System Environment*; ARL-TR-3815; Army Research Laboratory: Aberdeen Proving Ground, MD, 2008.

Dixon, S. R.; Wickens, C. D. Control of Multiple-UAVs: A Workload Analysis. *Proceedings of the 12th International Symposium on Aviation Psychology*, Dayton, OH, 2003.

Dixon, S. R.; Wickens, C. D.; Chang, D. Comparing quantitative model predictions to experimental data in multiple-UAV flight control. *Proceedings of the 47th Annual Meeting of the Human Factors and Ergonomics Society*, Santa Monica, CA, 2003.

FM 5-170, *Engineering Reconnaissance, 1998.* Headquarters Department of the Army: Washington D.C.

Fong, T.; Thorpe, C.; Baur, C. Multi-Robot Remote Driving with Collaborative Control. *IEEE Transactions on Industrial Electronics* 2003, *50* (4).

Hart, S. G.; Staveland, L. E. Development of NASA-TLX (task load index): Results of Empirical and Theoretical Research. In P. A. Hancock and N. Meshkati, Eds. *Human Mental Workload.* North-Holland: Elsevier Science Publishers, pp. 139–183, 1988.

Haselager, W.F.G. Robotics, Philosophy and the Problems of Autonomy. *Pragmatics and Cognition* 2005, 13 (3), 515–532.

Holm, S. A Simple Sequentially Rejective Multiple Test Procedure. *Scandinavian Journal of Statistics* 1979, *6* (2), 65–70.

Huang, H.; Pavek, K.; Albus, J.; Messina, E. Autonomy Levels for Unmanned Systems (ALFUS) framework: An Update. *Proceedings of the 2005 SPIE Defense and Security Symposium*, March 2005, Orlando, FL.

Lane, J. C.; Carignan, C. R.; Sullivan, B. R.; Akin, D. L.; Hunt, T.; Cohen, R. Effects of Timedelay on Telerobotic Control of Neutral Buoyancy Vehicles. *Proceedings of IEEE International Conference on Robotics and Automation*, 2002.

Luck, J. P.; McDermott, P. L.; Allender, L.; Russell, D. C. An Investigation of Real World Control of Robotic Assets Under Communication Latency. *Proceedings of the 1st ACM SIGCHI/SIGART Conference on Human-Robot Interaction*, 2006, pp. 202–209.

Maes, P. Artificial Life Meets Entertainment: Life Like Autonomous Agents. *Communications of the ACM* 1995, *38* (11), 108–114.

Malcolm, A. A.; Lim, J.S.G. (2003). Teleoperation control for a heavy-duty tracked vehicle. Retrieved June 21, 2004, from http://www.simtech.a-star.edu.sg/Research/TechnicalReports/TR0369.pdf.

Murphy, R. R. *Introduction to AI Robotics*. MIT Press: Cambridge, MA, 2000.

Schipani, S. An Evaluation of Operator Workload During Partially-Autonomous Vehicle Operations. *Proceedings of PerMIS 2003*, 16–18 September 2003, Washington, DC.

Scribner, D. R.; Dahn, D. *A Comparison of Soldier Performance in a Moving Command Vehicle Under Manned, Teleoperated, and Semi-autonomous*

Scalability of Robotic Controllers

Robotic Mine Detector System Control Mode; ARL-TR-4609; U.S. Army Research Laboratory: Aberdeen Proving Ground, MD, 2008.

Singer, P. W. *Wired for War: The Robotics Revolution and Conflict in the 21st Century*. The Penguin Press, NY, 2009.

Stubbs, K.; Hinds, P. J.; Wettergreen, D. Autonomy and Common Ground in Human-Robot Interaction: A Field Study. *IEEE Intelligent Systems*, March–April 2007, *22* (2), pp. 42–50.

APPENDIX A. OPERATIONS ORDER (ENGINEERING RECONNAISSANCE [FM 5-170], 1998)[*]

	Copy 1 of 10 copies HQ, 99th Engineer Battalion NK111111 080500 SEP 09

OPERATION ORDER 10-11

References:	1st Bde OPORD 10-23 Map sheet V107

Time Zone Used Throughout The Order: Local

Task Organization:

A/99 En Bn	B/99 En Bn 1/C/99 En Bn	Bn control C/99 En Bn (-) Recon Team 1

1. Situation

a. Enemy Forces

(1) Terrain and Weather

(a) Observation is generally limited along the valley floor due to the terrain's undulating nature. Multiple intervisibility lines, generally running

---

[*] This appendix appears in its original form without editorial change.

north to south and spaced between 500 to 1,000 meters, will hamper observation. Movement to the higher elevations along either the north or south wall will obviously improve observation. Winds (expected to exceed 20 knots until at least 111200 SEP 09) will lift sand from the desert floor and hamper observation. Observation at night will be extremely limited due to the light data for the next 72 hours. Note that on 8 SEP 09 the moon sets before the sun and on 9 through 11 SEP, the moon sets soon after the sun; therefore, night-vision goggles (NVGs) will provide limited capabilities for the next 72 hours and make observation, movement, and the acquisition of OBSTINTEL more difficult.

Date	BMNT	SR	SS	EENT	MR	MS	Start NVG	Stop NVG	% Illum
8 SEP	0555	0654	1701	1800	0550	1633	*****	*****	0%
9 SEP	0555	0653	1702	1801	0643	1743	*****	*****	0%
10 SEP	0555	0653	1702	1802	0731	1853	*****	*****	4%
11 SEP	0554	0653	1704	1803	0813	2002	*****	*****	9%

(b) The only cover from both direct and indirect fires is provided by the undulating terrain previously mentioned. Concealment during movement can be enhanced by traveling parallel to the intervisibility lines when available. The dusty and windy conditions may make mounted movement less detectable by the enemy.

(c) The pipeline running parallel to the LD along the 30 easting is the only existing obstacle in the AO. Crossing points for this pipeline have been identified at NK 302215 and NK 295090.

(d) The terrain in the vicinity of the templated obstacle system is believed to be unsuitable for minefield reduction by MCBs because of the undulating terrain and the soil composition.

(2) Enemy Situation

(a) The 133d motorized rifle battalion (MRB) is currently preparing defenses along the 47 easting. This unit's expected strength is estimated to be 12 T-80s, 32 BMP-1s, 3 AT-5s, and 1 dismounted infantry company. The 133d MRB began preparing its defenses 071500 SEP and are not expected to complete its counter mobility and survivability effort before 091600 SEP. The 133d MRB is expected to have a company-size combined-arms reserve at a strength of three T-80s and eight BMP-1s.

(b) As of 080100 SEP, three enemy MRCs have been located and are depicted on the SITEMP. The expected positioning of the subordinate MRPs is also templated as well as the anticipated combat security observation post (CSOP) and artillery positions. Expect to come within direct-fire range of the CSOPs when crossing the 42 easting and the main defenses when crossing the 44 easting. Enemy artillery is expected to be in position not later than (NLT) 081600 SEP; expect to come within indirect-fire range when crossing the 25 easting. However, the enemy will rarely use indirect fires against recon forces. Expect the enemy to use its rotary-wing assets in its attempt to try to locate and destroy recon forces. The enemy is not expected to be supported by fixed-wing aircraft. Although the enemy has the capability to employ chemical weapons, it has chosen not to do so thus far in the campaign. However, if the enemy does employ chemicals, we expect them to emplace a persistent chemical agent at center of mass NK 410280.

The enemy is not nuclear capable. Expect the enemy to use dismounted strong points to tie its obstacles into the restricted terrain at vicinities NK 450210 and NK 440100. These dismounted forces will be supplied with AT-5s to assist in their mission of preventing the obstacle system from being reduced along the walls of the valley. Additionally, the enemy will use dismounted patrols to protect all minefields.

(c) The templated obstacle system is included on the SITEMP. No confirmed obstacle locations have been obtained as of 080100 SEP. We expect the enemy to continue to lay its minefields similar to the method used throughout the campaign. We expect the enemy to mechanically lay its minefields and expect each minefield to be comprised of SB-MV mines and be 200 to 300 meters long and 60 to 120 meters deep. The mine spacing has consistently been 4.5 meters and the depths of the mines have been up to 9 inches.

NOTE: The SB-MV is magnetic-influence initiated and must be detected by probing. Operating hand-held mine detectors may detonate the mine.

If mines are surface laid, it is probably due to the soil conditions and indicates the probable success of using MCBs. The enemy has routinely used a single-strand of concertina fence on the enemy side of the minefield as a frat fence. The enemy is expected to emplace a total of 15 minefields in its defense.

b. Friendly Forces

(1) Higher.

(a) 1st Brigade plans to conduct a brigade breaching operation and penetrate the northern MRP of the northern MRC as shown on the SITEMP. Other brigade recon assets include two COLTs and one chemical recon vehicle. The planned locations for each of these assets are shown on the maneuver graphics .

(b) Engineer recon team 1 is attempting to answer the brigade commander's PIR for location, composition, and orientation of the enemy's obstacles.

(c) If bypasses of the enemy obstacles can be located, the brigade commander would prefer to bypass the obstacles as close to the north wall as possible.

(2) Lower. Do not expect TF recon assets to cross the LD before EENT on 9 SEP 09.

2. Mission

The 99th Engineer Battalion conducts an area recon of NAI 301 NLT 082000 SEP 09 to facilitate the brigade's attack at 110500 SEP 09.

3. Execution

Intent

The purpose of this mission is to identify enemy obstacles within NAI 301 to confirm or deny the enemy's COA and facilitate breaching operations. The end state is the identification of enemy obstacles in NAI 301 NLT 100500 SEP 09 and recon team 1 in position at checkpoint (CP) 15 ready to link up and guide the breach force to the obstacle location NLT 110001 SEP 09.

A. Concept of Operation

The battalion conducts an obstacle-oriented area recon. Recon team 1 will cross the LD NLT 082000 SEP. The brigade will have at least two batteries ready to provide indirect fires out to PL Celtics throughout the recon effort, and the attack helicopter battalion (AHB) will support casualty evacuation. Team 1 will cross the LD about 24 hours before the TF scouts in an attempt to observe and report the enemy emplaced obstacles and signs of recent enemy

Scalability of Robotic Controllers

activity while the TF is still planning its R & S effort. The recon team will link up with TF 1-23 scouts (who will provide security) before conducting obstacle recon. Recon team 1 will complete its area recon of NAI 301 NLT 100500 SEP to facilitate mounted rehearsals by the brigade during daylight hours on 11 SEP 09. Recon team 1 will continue to observe NAI 301 until 101700 SEP and report any further engineer activity. At 101800 SEP 09, recon team 1 will move to CP 15 and be in position NLT 110001 SEP 09, prepared to link up and guide the breach force to the obstacle location.

B. Tasks to Subordinate Units

(1) Battalion's TOC. The battalion's TOC will--

(a) Provide liaison personnel to co-locate with the recon team until they cross the LD and ensure that liaison personnel obtain a copy of the recon team's maneuver graphics.

(b) Coordinate the recon team's indirect fire plan with the FSO and confirm targets with the team leader once they are coordinated.

(2) Battalion S4. The battalion S4 will obtain the current logistical status of recon team 1. He will ensure that unit basic load (UBL) levels are reestablished NLT 081200 SEP 09 and report to the TOC upon completion.

(3) A/99 En Bn. A/99 En Bn will conduct liaison activities between recon team 1 and TF 1-23 according to the battalion's TACSOP.

(4) Recon team 1. Recon team 1 will--

(a) Report the current logistical status to the S4 NLT 080800 SEP 09.

(b) Backbrief the plan to the battalion commander via FM radio at 081300 SEP 09.

(c) Provide TF 1-23 the team's graphics, via the A/99 En Bn's TOC before crossing the LD.

(d) Forward requested indirect-fire targets to the battalion's TOC NLT 081600 SEP 09.

(e) Coordinate link up with the TF 1-23 scouts for security during obstacle recon.

(f) Conduct an area recon of NAI 301 NLT 082000 SEP 09 to verify the composition of obstacles within the NAI.

C. Coordinating Instructions

(1) Task organization is effective upon receipt of this order.

(2) All units will participate in the intelligence updates to occur at 0800 and 2000 each day.

(3) The LOA for recon assets is PL Celtics.

4. Service Support

A. Support Concept

(1) The recon team will cross the LD fully uploaded according to the battalion's TACSOP. These supplies will come from the engineer battalion. This basic load is expected to sustain the team throughout the mission.

(2) Emergency resupply will be coordinated through the engineer battalion's TOC and delivered by aviation assets. Backup resupply will be through TF 1-23.

b. Medical Evacuation and Hospitalization. The primary means of MEDEVAC is by air (requested through the battalion's TOC); backup is by ground evacuation (performed by TF 1-23).

c. Personnel Support. EPWs will be turned over to TF 1-23 for evacuation to the rear.

5. Command nd Signal

A. Command

The chain of command is the commander, the XO, the S3, and the commander of C Company.

B. Signal

(1) All traffic from recon team 1 to the battalion's TOC will be over MSRT (primary) or the battalion's command net (alternate).

(2) The recon team's current location will be sent by the battalion's TOC to TF 1-23.

(3) OBSTINTEL will be reported according to the TACSOP.

ACKNOWLEDGE:

PATTON

LTC

APPENDIX B. DEMOGRAPHICS[*]

SAMPLE SIZE = 30

	MOS	RANK	DUTY POSITION	
09S – 10	68D – 1	E4 – 6	Air Aslt Instructor – 1	Pathfinder Instructor - 1
11B – 8	68W – 1	E5 – 17	Crew Chief – 1	Ranger Instructor – 1
11C – 1	88M – 2	E6 – 2	Driver – 1	Support – 1
12B – 1	0CS – 3	OCS – 4	Instructor – 1	Team leader – 4
15T – 1	NR – 1	NR – 1	Medic – 1	Training – 1
31B – 1			OCS – 7	NR - 9
			Combat Engr - 1	

Age

28 years (range 21-47)

1) How long have you served in the military? <u>65 </u>months (mean)
2) How long have you had an infantry-related job? <u>72 </u>months (mean)
3) How long have you been a fire team leader? <u>21 </u>months (mean)
4) How long have you been a squad leader? <u>44 </u>months (mean)
5) How long have you been deployed overseas? <u>21 </u>months (mean)
6) How long have you been deployed in a combat area? <u>15 </u>months (mean)
7) With which hand do you most often write? <u>25 </u>Right <u>5 </u>Left
8) With which hand do you most often fire a weapon? <u>27 </u>Right <u>3 </u>Left
9) Do you wear prescription lenses? <u>4 </u>Yes <u>26 </u>No
10) If yes, which do you wear most often? <u>2 </u>Glasses <u>2 </u>Contacts
11) Which is your dominant eye? <u>26 </u>Right <u>4 </u>Left
12) Do you have any vision related problem? <u>2 </u>Yes <u>27 </u>No <u>1 </u>NR If so, what? Red/green color blind (1), farsighted (1)
13) Have you ever used a robotic system? <u>1 </u>Yes <u>23 </u>No <u>6 </u>NR If so, what type? Davinci (1)
14) Please describe the conditions under which you used the robotic system. Surgical (1)

---

[*] This appendix appears in its original form without editorial change.

15) Using the scale below, please rate your skill level for each of the following activities.

None	Beginner	Intermediate	Expert
1	2	3	4

ACTIVITY	MEAN RESPONSE
Operating ground unmanned vehicles	1.13
Operating aerial vehicles	1.09
Target detection and identification	1.55
Playing commercial video games	2.65
Training with Army video simulations	2.17

APPENDIX C. TRAINING QUESTIONNAIRE[*]

SAMPLE SIZE = 30

1. Using the scale below, please rate the training you received in the following areas.

1	2	3	4	5	6	7
Extremely bad	Very bad	Bad	Neutral	Good	Very good	Extremely good

	MEAN RESPONSE
a. Introductory robotic training	
--Completeness of introductory training	5.90
--Comprehension of the overall concept of the robot	5.90
b. Building reconnaissance training	
Robotic driving	
	MEAN RESPONSE
--How to drive the robot (teleoperation)	5.73
--How to drive the robot (semiautonomous)	5.70
--How to drive the robot (autonomous)	5.97
Time provided to practice driving the robot on the building reconnaissance course	
--Teleoperation	5.59

---

[*] This appendix appears in its original form without editorial change.

Scalability of Robotic Controllers

--Semiautonomous	5.55
--Autonomous	5.81
How to complete the building reconnaissance course (independent of levels of autonomy)	
c. Overall evaluation of the robotic training courses	5.96

Comments	No. of Responses
I personally was challenged; not a video gamer.	1
Instructions were great!	1
b. Building Reconnaissance Training I liked the tele mode best.	1
More time makes you better, but for what we did today, it was enough.	1
The auto was the easiest level because it gives you the opportunity to multitask a lot easier than the other two settings.	1
More distractions perhaps.	1
Attempting to draw to scale, take pictures, and maintain awareness may be a problem in the tactical field.	1
Drawing/mapping was difficult.	1
c. Overall Robotic Training	
Extremely fun and helpful for the future of operation.	1
Great ideas, tech, instruction.	1
d. Speech Training	
Would free up more time and frustration.	1
Need a little more time with it. I felt like I had more time with grenades and the robot besides the speech control.	1
Not good at all in my case, but that has to do with my ability in speaking English.	1

2. What were the easiest and hardest training tasks to learn?

Comments	No. of Responses
Easiest	
All of them.	1
Training on all was performed well.	1
The task of each control.	1
Control of robot (driving, steering, moving, maneuvering, etc.).	12
Use of controller.	1
Using controls to draw the map.	1

Appendix C. (Continued)

Comments	No. of Responses
Learning how the remote control works, and what buttons do what.	2
Taking pictures.	8
Collect intel.	2
Photographing and identification.	1
Identify weapons and other items.	1
Labeling pictures.	3
Identifying obstacles.	1
Autonomous was the easiest.	2
Tele and Semi were the easiest.	1
Hardest	
None of them, at least to me.	1
Doing the entire job. Better to just be in the room.	1
Feeling the turn radius of robot.	1
Getting robot to move easier. They should find a way to have pivot points instead of tracks so the robot eye is multidirectional and independent of the mobility of the robot itself.	1
Robot has difficulty with full autonomous mode in close areas.	1
Prioritizing the video with the map for situational awareness.	1
How to work my way through the building. I didn't like the delay using the map in relation to my controls was unclear.	1
Mapping the building by hand.	4
Drawing the mapduring building reconnaissance.	1
Answering questions while driving robot.	3
Situational awareness.	1
Object identification.	1
Driving around obstacles.	1
Speech training/controls.	2
Taking pictures. But once you get used to it, it becomes very easy.	1
Labeling the pictures.	1
Overall functions of each button.	2
Using the joystick/control pads.	1

Understanding that you have "tunnel vision" with the robot.	1
Tele was the hardest.	2
Autonomous was the hardest.	1

3. What are your overall comments on the training?

Auto and Semi worked great.	1
Good.	7
Good learning experience; something I've never seen before.	2
Fantastic tool. Can see how this will save more lives by using the robot as the recon tool.	1
Interesting and fun.	3
Easily understood what to do during these exercise.	1
Fairly simple.	1
Cadre was very informative.	1
I'm glad I got to participate.	1
Would like to see how it works with a two-man team with the same evaluation.	1
I feel everyone was very knowledgeable and professional.	1
Full instructions; good review.	1
I felt very comfortable operating the robot.	1
Could use on battlefield in urban areas.	1
I would like to do it in a real mission.	1
Could use more time to practice controlling robot.	1
Driving the robot and not running into everything, like the edge of a door.	1
Display hard to see.	1
The use of the directional pad would have been great to have known about in the training so I could practice.	1
The tele operation and map drawing was extremely difficult to do both under a set time.	1

Appendix D. Post Iteration Questionnaire[*]

SAMPLE SIZE = 30
AUTONOMY LEVEL:
TELEOPERATION (TELE), SEMIAUTONOMOUS (SEMI),
AUTONOMOUS (AUTO)

1. Using the scale below, please rate your ability to perform each of the following **tasks** based on your experience with the autonomy level that you just used.

1	2	3	4	5	6	7
Extremely difficult	Very difficult	Difficult	Neutral	Easy	Very easy	Extremely easy

	MEAN RESPONSE		
	TELE	SEMI	AUTO
a. Move the robot in the correct direction	5.23	5.20	6.00
b. Avoid obstacles	5.14	5.50	5.48
c. Avoid pot holes	5.67	5.50	7.00
d. Assess down slopes for navigatability	5.50	5.67	7.00
e. Assess side slopes for navigatability	5.50	5.67	7.00
f. Identify any other terrain features that might have an adverse effect on the ability of the robot to	6.13	6.38	6.00
g. Anticipate whether the ground clearance of the vehicle will allow negotiation of rugged terrain	6.25	6.80	6.67
h. Anticipate whether the turn radius of the vehicle will allow a turn	4.67	4.65	5.32
i. Identify if the robot is on the correct path	5.38	5.37	5.72
j. Navigate far enough ahead to plan route in	5.14	5.21	5.27
k. Navigate well enough to drive at slowest speeds	6.00	6.17	6.22
l. Navigate well enough to drive at medium speeds	5.75	6.00	6.05
m. Navigate well enough to drive at fastest speeds	5.32	5.38	5.74
n. Finish the course quickly	4.07	4.75	5.34
o. Ability to find IEDs and other objects of interest	5.69	5.86	6.00
p. Ability to navigate to the next waypoint	5.50	5.68	5.72

---

[*] This appendix appears in its original form without editorial change.

	MEAN RESPONSE		
	TELE	SEMI	AUTO
q. Ability to map the room (building reconnaissance course only)	3.75	6.00	6.21
r. Ability to take pictures	6.40	6.29	6.37
s. Ability to make the robot understand you	5.88	5.21	5.65
t. Ability to maintain situation awareness	4.37	4.97	5.38
u. Overall ability to perform this reconnaissance with this level of autonomy	4.35	5.20	5.86

Comments	No. of Responses
TELE	
Mapping room accurately was difficult.	1
If the camera on the robot could zoom in and out, it would help out a lot.	1
Camera does not allow any view other than forward at ground level. Can't tell how close it is to objects. Can't see anything that could be happening behind robot, i.e., rear camera.	1
This mode is difficult as one person. If a two-man team was involved, it would be easier by a lot.	1
SEMI	
Seems the easiest and most effective.	1
Good, because you can let the robot move on its own and stop it and take control if you wanted to.	1
Performing functions on the controller became easy to remember after a few repetitions; approx 10 minutes.	1
The directional pad wouldn't respond well (slow or not at all). The robot moves extremely slow and "thinks" too much for each breach of a room. It was faster to just let it go into a room and then take control.	1
The joystick sensitivity is not high enough.	1
In my opinion, this level is worthless because I had to operate it more than by itself.	1
I prefer the autonomous more.	1
AUTO	
It was easy for me to answer questions and look up info in this mode.	1
Easier with practice.	1

Appendix D. (Continued)

Comments	No. of Responses
Overall felt like the same as semi. Seemed more efficient to continually take over rather than allow robot to maneuver on its own the whole time due to number of items.	1
Unable to label some objects dueto low camera resolution or perhaps light levels.	1

2. Please check any of the following conditions that you may have experienced during this trial.

	NUMBER OF RESPONSES		
	TELE	SEMI	AUTO
Eyestrain	7	7	4
Tunnel vision	2	2	1
Headaches	0	1	0
Motion sickness	0	0	0
Nausea	0	0	0
Disorientation	2	0	0
Dizziness	0	0	0
Competition between eyes for vision of different scenes at which they are looking	2	3	1
Any other problems?	0	1	0

Comments	No. of Responses
BUILDING RECONNAISSANCE	
TELE	
Competitionbetween eyeswhen trying to look at the screen and draw at the same time on your map. If I were the operator, I would have an assistant to map it.	1
Eyestrain from glare on the computer screen.	1
Screen should be bigger or be in room with less light.	1
SEMI	
Some items were a little hard to make out with the cameraand caused eyestrain.	1

Comments	No. of Responses
AUTO	
Training is easy.	1
Some items were a little hard to make out with the cameraand caused eyestrain.	1
I would be able to ID smaller objectives if there was a zoom feature on the camera.	1
Possible color vision.	1

3. Using the scale below, what is your **overall rating** of the level of autonomy you used on this course.

1	2	3	4	5	6	7
Extremely bad	Very bad	Bad	Neutral	Good	Very good	Extremely good

MEAN RESPONSE		
TELE	SEMI	AUTO
4.63	5.11	5.63

Comments	No. of Responses
BUILDING RECONNAISSANCE	
TELE	
Useful and probably more practical; however, it is difficult to do as one person.	1
I prefer the control of the robot, but ability to map was difficult.	1
Having to drive, map, and answer questions was too much.	1
When in full tele mode, it is hard to concentrate on main mission while having to react to outside influences. With this, frustration comes in making tasks seem even more difficult.	1
The controls are hard to turn.	1
Fixed camera may be a problem.	1
Rear camera would be nice as well.	1
Robot if tactically possible could sit higher from ground level or the camera.	1

Appendix D. (Continued)

SEMI	
Allowed focus to be shifted to the secondary objective while maintaining the primary objective.	1
Comments	No. of Responses
Like playing with a toy/Xbox.	1
I had a tendency to want to have more control over the robot.	1
Rough control over the robot.	1
I didn't like it at all. I'd rather use Tele where I am in total control. But, of course, I believe that would be subject to change if I had more practice with it.	1
Robot overlooked nearly an entire room, and got itself stuck against a wall. Definitely had to navigate some with remote; very simple.	1
This level is worthless to me and the camera could be better with clarity and the ability to zoom in and out.	1
Not as aware of surroundings.	1
AUTO	
Robot did its job very well allowing me to add focus to secondary objectives without compromising my primary goal.	1
This mode makes it easier than the other two. It allows you to multitask a lot easier than the other two.	1
Autonomous mode seems to be the way to go.	1
Very good. Can I come back and do this tomorrow?	1
I think the camera should be providing a better resolution and should be moveable.	1

APPENDIX E. END OF EXPERIMENT QUESTIONNAIRE[*]

SAMPLE SIZE = 30

1. Using the scale below, please rate the following characteristics of the control system that you used.

1	2	3	4	5	6	7
Extremely bad	Very bad	Bad	Neutral	Good	Very good	Extremely good

---

[*] This appendix appears in its original form without editorial change.

Scalability of Robotic Controllers

CHARACTERISTICS	MEAN RESPONSE
a. Resolution (clarity) of the display	4.63
b. Size of objects appearing in the display	4.90
c. Ability to adjust display	3.83
d. Comfort of viewing the display	4.30
e. Display brightness	4.77
f. Effect of glare on the display	4.03
g. Contrast between objects on the driving display	4.93
h. Display color	4.13
i. Comfort of using the display	4.73
j. Number of controls	5.23
k. Control locations	5.43
l. Size of individual controls	5.67
m. Complexity of controls	5.40
n. Ability to use controls without inadvertently activating other controls	5.60
o. Size of entire control unit	5.80
p. Adequacy of this control unit for teleoperating a robot	5.60
q. Adequacy of this control unit for semi-autonomously operating a robot	5.67
r. Adequacy of this control unit for autonomously operating a robot	5.87
s. Overall assessment of this control unit	5.40

2. Please indicate your 1st, 2nd and 3rd choice of the level of autonomy to use for completing the following tasks by placing a 1, 2, or 3 in the appropriate column

	MEAN RATING		
	Tele-operation	Semiautonomous operation	Autonomous operation
Route planning	1.62	2.04	2.35
Generating goals	1.88	1.96	2.20
Driving a straight route	1.96	2.15	1.85
Avoiding obstacles	1.81	2.04	2.15
Looking for objects of interest	2.38	1.96	1.81
Overall building reconnaissance	1.73	1.96	2.31

Appendix E. (Continued)

3. What suggestions do you have for ways to increase the effectiveness of the following modes?

Comments	No. of Responses
Teleoperation	
I liked tele because I was in control; but I wasn't a fan of the mapping.	1
Tele mode offers more freedom and choice when operating the robot, but it takes too much time.	1
Easy to operate but you have to do everything manual so it is a little more time consuming.	1
Side/rear cameras.	1
Robot still has the ability to map.	1
Symbols on control pad to show what it will do (photos).	1
Improve lag.	3
Adjust screen so that visibility is clearer.	1
Wide angle screen.	1
Being able to use the laser rangefinder.	1
Better zoom on camera.	1
Better camera clarity and angle of view.	1
Hot keys to quickly mark/label items of interest.	1
Color monitor.	1
Fewer distractions.	1
Automatically go to label a picture after one is taken.	1
The controls for picture taking and being able to label were confusing	1
after a while.	
Better nav controls.	1
Improve signal for response.	2
Less questions.	1
Semiautonomous	
Works well.	1
More responsive to user suggestions while driving.	1
Responsiveness could be slow, from switching in and out of semi mode. Got stuck at one point.	1
Make more like autonomous where it will scan the room while moving.	1
Robot identifies objects that are inconsistent with surroundings more clearly.	1

Use to ID objects.	1
Lower speed to stop for pictures.	1
Comments	No. of Responses
Improve lag.	2
Better zoom on camera.	1
Hot keys to quickly mark/label items of interest.	1
Get rid of it because you spend more time in the tele mod.	1
Color monitor.	1
Wide angle screen.	1
Did not like because it was like a retard robot.	1
Symbols on control pad to show what it will do (photos).	1
Could have a temp pause button instead of two (one to stop and one to resume).	1
Autonomous	
This is the best mode.	1
Worked great!	1
Excellent for the job and really simple.	1
Works well.	1
Speedy.	1
Full room scans via camera.	1
Color monitor.	1
Wide angle screen.	1
Better zoom on camera.	1
Hot keys to quickly mark/label items of interest.	1
Screen clarity.	1
Make sure the robot travels to every corner of the room so the operator can see everything before it moves on to the next room.	1
Limits the effectiveness of how well you can conduct recon on a building.	1
Symbols on control pad to show what it will do (photos).	1
Help robot navigate around objects easier.	1
Time delay.	1
More practice on it.	1

4. Using the scale below, please rate the level of reliability (trust) of the autonomous systems you used.

1	2	3	4	5	6	7
Extremely unreliable	Very unreliable	Unreliable	Neutral	Reliable	Very reliable	Extremely reliable

No comments.

Appendix E. (Continued)

5. Using the scale below, please rate how helpful the following capabilities were to your ability to drive/control the robot in the semi-autonomous and autonomous modes.

1	2	3	4	5	6	7
Extremely unhelpful	Very unhelpful	unhelpful	neutral	helpful	Very helpful	Extremely helpful

CAPABILITY	MEAN RESPONSE
a. Robot's position/location	5.57
b. Showing the path the robot drove	5.80
c. Obstacle avoidance	5.53
d. Ability to return to "start location"	6.27
e. Map data (robot built)	6.11
f. Capability of the robot to drive itself	6.00

6. Would you want the following capabilities again if you have to perform these same missions over again?

	NUMBER OF RESPONSES		
	YES	NO	NR
a. Robot's position/location	25	1	4
b. Showing the path the robot drove	25	1	4
c. Obstacle avoidance	26	0	4
d. Ability to return to "start location"	25	1	4
e. Map data (robot built)	24	0	6
f. Capability of the robot to drive itself	26	0	4

7. How would you improve the robot and/or any of its features?

Comments	No. of Responses
Better camera with higher resolution.	7
Improve clarity or provide light source of some kind.	2
Bigger camera view.	3

Camera positioning and control.	1
Moveable camera with zoom feature.	2
Comments	No. of Responses
Add a wider lens that elevates and swivels for easier identification.	1
An arm and elevation on camera.	1
Add a scanning laser that identifies objects as question marks on the screen in the auto mode.	1
Constant stopping and adjusting view in the tele mode for pictures took lots of time. Maybe add in a "room scan" so the robot automatically visually covers whole room.	1
Photos can be pop-up windows (1-10), then you can label from there instead of having a picture in the middle of your pathway on the display.	1
Adjust photo movement (up and down).	1
If the robot would recognize and take photos on its own.	1
Add an extension for the lens.	1
Lag time was main problem and it got confused sometimes concerningwhere it was going; maybe make an adjustment for that.	1
Avoiding obstacles a little better in semi and auto modes.	1
The avoidance feature needs to be able to be deactivated.	1
Map data needs to be clearer.	1
Improve the elevation view and the capability to climb stairs and not get stuck.	1
Height of robot.	1
Speed.	1
Turning ability.	1
Make it lighter so it would be easier to transportand add a light to illuminate objects that are in the dark.	1
Quieter.	1

8. How would you improve the interface and/or controller?

Comments	No. of Responses
Already good.	2
Very easy to understand; leave as is.	5
Most soldiers are into games, so that type of controller works in this situation.	1
Perfect controller.	2
Different control.	1

Appendix E. (Continued)

	No. of Responses
Add hot keys to quickly mark/tag items of interest.	1
Comments	No. of Responses
Add a button for elevation scan.	1
Add a warning LED when in tele to warn the user of objects that theywill hit.	1
Add an instrument panel that shows distance cleared, total distance traveled, height of the room, etc.	1
Improve lag on display.	1
Identify objects on map more clearly.	1
Video display should take up about half of the screen. When a lot of objects were nearby, it was difficult to tell one picture from another. The labels should be in alphabetical order when they are displayed.	1
Faster mouse speed.	1
Asking for a label automatically whenever a picture is taken.	2
The X,Y,A,B buttons would be better for selecting items, taking pictures, zoom in and out. These are the closest buttons to your hand and the most used. The LB and RB are the least used in Xbox games for that reason. They are inconveniently placed and awkward.	1
Alternatecontrols (dual-stick tank control).	1
It is easier to view with right thumb stick left/right and move with the left thumb stick.	1
Perhaps 360oviewing ability.	2
Get more input from gamer for functionality.	1
Bigger joystick with movement buttons on it instead of controller front.	1
Larger screen for room/path image.	1
Wide angle lens.	1
Color.	1
Zoom feature on the video screen bottom; sort of like a drop-down menu when you change the size of your pictures on your PC when you print.	1

9. a. Did you feel a difference between the semi-autonomous and autonomous modes?

9 No
21 Yes
b. If yes, what was the difference(s)?

Scalability of Robotic Controllers

Comments	No. of Responses
Auto	
The auto mode will go into a room and scan the room for you giving you more time tolook for things.	1
Lets you go from A to B and work backwards from B to A.	1
Seems to return to the starting position more quickly and efficiently.	1
Ease of operation and speed.	2
More aware.	1
Maps better.	2
Easier to focus on what to take photos of, instead of doing both.	1
Total tech control.	1
Recons the site faster.	1
Semi	
Semi mode moved distances faster.	1
The robot moving itself.	1
Ease of movement in between objects.	1
Semi seemed to be smoother with the controls.	1
More control (without stopping).	3
Obstacle avoidance and new unidentified exploring.	1
User input during semi mode was helpful.	1
You pay attention to making corrections and having to over compensate.	1
Messed up (got stuck, lagged —not a fan!).	1
Does not seem as aware as auto.	1
General	
The difference was hard to discern because I was stopping the robot just as often in both to take pictures.	

10. a. Which mode do you prefer?

<u>16</u> Autonomous

<u>14</u> Semi-autonomous

b. What are the reasons for your preference?

Comments	No. of Responses
Autonomous Didn't lag, ran smoother. I never got stuck. Felt like I could look at OPORDER and not worry if it'd get stuck or confused.	1
Easier to work with because it goes to every corner of the building.	1

Appendix E. (Continued)

Comments	No. of Responses
Easy to operate and I like to see the area with everything in it.	1
More comprehensive and fully maps building.	1
Ability to focus more on the screen than controlling.	1
Allows me to concentrate on objects that can make/break the mission success.	1
Ease of operation and speed.	1
Lets you multitask easier.	3
Seemed to know where it was going.	1
I don't really get the difference between the two. I understand what it's supposed to do, but mostly I just stopped it and ran it on my own.	1
Semi-autonomous	
Ability to guide robot during travel.	1
It still seemed necessary due to the number of items and also the robot started to skip past a room. Although it may have eventually gone	1
back, it seemed more efficient to just take control.	
Semi is faster at distance travel.	1
I like the interactive control.	1
I feel I have more and faster control of the robot.	5
Moving on its own and still having control.	1
Easy to change course slightly.	1
Best feature is the robot driving itself. Therefore, they were both a good feature.	1
Exploring something quickly and laying out the floor plans.	1
Manual control is preferred.	1

11. In the semi-autonomous and fully-autonomous modes, the robot automatically built a floor plan of the building for you. How often did you use the floor plan as you looked for objects of interest? How was it helpful or not helpful?

Comments	No. of Responses
Very often. Absolutely helpful.	7
It helped me know where I was at all times.	1
I constantly viewed the mapped floor planto know where the robot had gone and where it needed to go.	1
Allowed for easy positioning, along with mission check.	1

Comments	No. of Responses
Map reading and using a map may seem out of date; however, it is of great importance to the Infantry.	1
Probably the most important tool. I used it to see where the robot had been and what pictures I'd taken, areas I needed to search, and the route I took. It was a very good and important asset to the operation.	1
I used the floor plan just to see the parameters of the building andto see where the robot was in the building initially. But I could have figured these things out myself. However, the floor plan was helpful because it was an added visual that showed where the robot had been in the building during the recon.	1
80 percent of the time. Allowed you to scan for items or interest.	1
Used it often to find out where I was and where I needed to look.	2
Used often to check for missed doors.	1
It was helpful, but didn't use it very much.	1
It helped more with the basic area to work in.	1
You knew where you have already been as far as room clearing. It was helpful in that manner.	2
Good for reference to whether photographed object yet or not.	1
I used it to make sure I didn't miss anything and rooms. When I was in full control, I worked the rooms clockwise making sure it was completely cleared before moving on, and then continued. When done, I went straight back to the beginning by pressing the return home button.	1
I didn't really use it other than to tell if I had already been in that room. Semi-helpful.	1
Somewhat helpful. The area being explored was small, so the mapping feature was less important. However, in a large building the mapping feature would be extremely useful.	1
It was helpful until it started wandering aimlessly without completing the course. The floor plan confused me.	1
It was helpful but not fresh in the room. Maybe to have an understanding for other sources or reasons. For example, the robot takes pictures and has a floor plan. Now another group can come in to collect intel knowing already what's there getting themselves in it.	1
Not very helpful at first, but as I got more comfortable, the usefulness increased.	1
Sometimes I spent more time looking at/for objects and determining what they were.	1

Appendix E. (Continued)

Comments	No. of Responses
I did not use it much to find objects; however, the map is incredibly useful for site recon and future operations. It gives a much more accurate assessment than I could draw using the display.	1
I only used the floor plan to go back and make sure I didn't miss any hidden items.	1

LIST OF SYMBOLS, ABBREVIATIONS AND ACRONYMS

ACS	Autonomous Capability Suite
ANOVA	analysis of variance
ARL	U.S. Army Research Laboratory
COTS	commercial grade off-the-shelf
GPS	global positioning systems
HRED	Human Research and Engineering Directorate
IMU	inertial measurement unit
MOCU	Multi-Robot Operator Control Unit
MOUT	military operations in urban terrain
NASA-TLX	National Aeronautics and Space Administration-Task Load Index
OCS	Officer Candidate School
POW	prisoner of war
SA	situation awareness
SSC Pacific	Space and Naval Warfare Systems Center San Diego
SUGV	small unmanned ground vehicle
WTC	Warrior Training Center

In: Robotic Autonomy and Control
Editors: T.C. Mueller and M.E. Bynes

ISBN: 978-1-62100-605-3
©2012 Nova Science Publishers, Inc.

Chapter 3

INTUITIVE SPEECH-BASED ROBOTIC CONTROL[*]

Elizabeth S. Redden, Christian B. Carstens and Rodger A. Pettitt

1. INTRODUCTION

1.1. Statement of the Problem

Two previous studies conducted by the U.S. Army Research Laboratory's (ARL) Human Research and Engineering Directorate (HRED) indicated that the degree of effectiveness of speech-based control may be dependent upon the task being performed (Cassenti, Kelley, Swoboda, and Patton, in press; Pettitt, Redden, and Carstens, 2009). The goal of the present research was to examine the effectiveness of speech control for the specific tasks used in robotic reconnaissance missions.

---

[*] This is an edited, reformatted and augmented version of an Army Research Laboratory publication, Aberdeen Proving Ground, MD 21005-5425, ARL-TR-5175, dated April 2010.

1.2. Robotic Control

As robots become more complex and able to perform multiple simultaneous tasks, menus become more layered, and speech recognition systems become more sophisticated, verbal control of robots appears to offer great potential to bridge the gap between teleoperation and full autonomy. An important consideration for optimal design is the identification of tasks that are more efficiently performed by speech control and those that are more efficiently performed by manual control.

Conventionally, robots are controlled with manual input devices such as joysticks, touch screens, and trackballs. While these devices are ubiquitous and have been used in countless applications, they come with their own set of problems when used to control robots in operational settings. Some robots are large; some require operation from stationary positions; many require dexterity, hand-eye coordination, significant training, and practice time; and all require the use of at least one hand. Alternatively, speech control has been demonstrated to provide many benefits. The following list of strengths associated with speech control is representative of those found in the literature (Chen, Hass, Pillalamarri, and Jacobson, 2006; Bortolussi and Vidulich, 1991; Steeneken, 1996; Bourakov and Bordetsky, 2009; Graham and Carter, 2006; Pettitt, Redden, and Carstens, 2009; and Sams, 2009):

- Is more effective than manual control for menu navigation (does not require navigation through several menu layers to access the desired item).
- Is quicker and more accurate than manual control for selection of options.
- Enhances time-sharing efficiency when used in conjunction with manual controls.
- Is hands free and eyes free.
- Reduces adverse effects of mobile device operation on primary task performance (i.e., driving, walking, etc.).
- Is effective for performance of simultaneous tasks (e.g., lowering the robotic arm while driving forward).
- Is intuitive if commands are tailored to the target audience.

Since speech-control systems are in their infancy, they also come with problems. The following is a representative list of weaknesses of speech control found in the literature (Chen et al., 2006; Noyes, Baber, and Leggatt,

2000; Myers and Cowan, 2003; Steeneken, 1996; Henry, Mermagen, and Letowski, 2005; Vloeberghs, Verlinde, Swail, Steeneken, and South, 2000; Graham and Carter, 2006; Cassenti, Kelley, Swoboda, and Patton, in press; Pettitt, Redden and Carstens, 2009; and Sams, 2009):

- Speech control is often infeasible in environments found in the military (i.e., stealth missions; high noise, high g-force, and vibration environments; impulse and variable noise environments; environments in which echoes, reverberation, and cross-talk are present; etc.).
- Communication between team members can sometimes be misinterpreted as voice commands.
- Speaker dependent speech recognizers require long training periods when the vocabulary consists of hundreds of words, while speaker independent speech recognizers generally have lower accuracy than speaker dependent ones.
- Speaker changes caused by the Lombard effect (the increase in vocal effort when a speaker is in a noisy environment), stress, fear, sickness, whispering, pain, "incorrect" use of grammar, wearing oxygen and gas masks, etc., cause word error rates to increase.
- Explicit feedback of recognition results has been found to be necessary.
- Speech control is not as effective as manual control for continuous tasks.

Speech control has been explored for many applications. For example, it has been examined for controlling helicopters, executing telephone and automobile functions, automating the handling of customer calls, converting spoken language into sign language, converting and analyzing large volumes of spoken material, performing computer command and query, translating or summarizing information from one language to another, creating medical records, and editing (Chen et al., 2006; Steeneken, 1996; Apaydin, 2002). In situations where equipment operators have busy eyes and hands, offloading control tasks to speech has been shown to be effective (Steeneken 1996). In order to control a robot with speech commands, the system must first be capable of recognizing the command. Automated speech recognition (ASR) systems digitize spoken words and match them against coded dictionaries. Once they are identified, the spoken information can control the actions of a system or machine (Haas and Edworthy, 2002).

Karis and Dobroth (1995) proposed that a successful human factors design of a speech recognition system should involve an early focus on the target audience for the system and the tasks they will perform, and collect performance data via simulations. Our experiment was an attempt to do just that.

The following are our hypotheses regarding speech control:

1) Speech control will be quicker than manual control when the operator must perform another task in addition to robotic control.
2) Speech control will be quicker than manual control when operators must access embedded menu items.
3) Speech control will be slower than manual control for performing continuous tasks, such as turning during a driving task.

In this study, we attempted to isolate the contribution of speech control to varying tasks associated with a robotic reconnaissance mission under differing conditions.

This experiment took place at Fort Benning, GA. Twenty-nine Soldiers from the Officer Candidate School (OCS) and instructors from the Warrior Training Center participated in the study.

2. METHOD

2.1. Participants

The twenty-nine Soldiers recruited from the OCS and the Warrior Training Center to participate in the study ranged in age from 21 to 47 years and had a mean time in the military of 65 months.

2.2. Apparatus and Instruments

2.2.1. Apparatus

Speech Control System

The SPEARTM speech control technology used in this experiment was developed by Think-AMove (TAM), Ltd. The SPEAR earpiece (figure 1) has a microphone inserted into the ear canal that captures the speech signal. A wired connection carries the signal from the earpiece and ends in a standard 3.5-mm audio jack that can be plugged into a computer soundcard. Through the design of the earpiece, the signal in the desired frequency range can be amplified, thus restoring the quality of the captured speech. A proprietary speech command recognition (SCR) system is used to identify the command spoken by the user. The SCR system collects the captured speech signal and sends out a recognized command. TAM has trained specialized models tuned for recognition of in-ear speech and the recognition accuracy target has been set to more than 90% even in extreme conditions, which include battlefield conditions where the operator is involved in intense physical activity, with loud noise in the background.

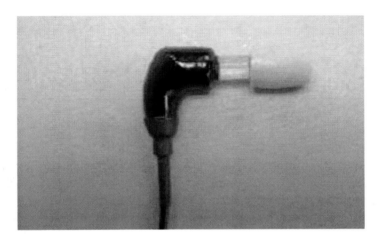

Figure 1. SPEAR earpiece.

Operator Interface

The operator interface that controlled the simulated robot used during this experiment was based on SSC Pacific's Multi-Robot Operator Control Unit (MOCU) and TAM's speech command library. An example screenshot of the interface is found in figure 2. The robot's location, driven path, goal points, and sensor data (i.e., map data) are overlaid on an aerial image. Various button controls and a joystick on a commercial-off-the-shelf (COTS) wireless game controller (the Microsoft Xbox 360 wireless controller) were used for manual control of the robot. A speech window was also provided so the operator could verify the accuracy of the speech command recognition.

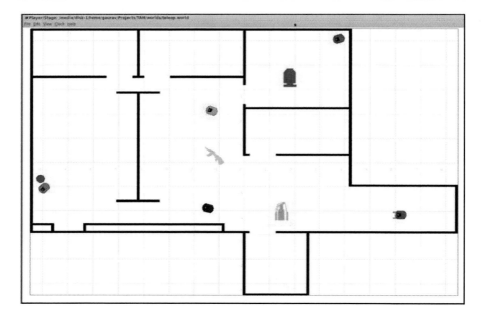

Figure 2. Snapshot of MOCU interface.

2.2.2. Instruments Control Intuitiveness Test

The control intuitiveness test that was performed in this experiment is roughly based on an icon intuitiveness study done by Nielsen and Sano (1994), which was performed during the usability evaluation of Sun Microsystems' internal Web. In the study, images were presented to users who were asked to indicate what functionality they thought the icons represented. In a like fashion, the control intuitiveness test presents a task to naive users and then asks them to get the robot to perform the task using speech when they have had no prior training or knowledge of the correct word or phrase to use. The administrator documents the incorrect attempts to perform the task and notes the specific words that are incorrectly used. After this attempt, the administrator provides them a list of correct commands and then asks them to perform the same tasks that were presented in the initial presentation. The administrator again notes the incorrect attempts to perform each task and notes the specific words that are incorrectly used.

Questionnaires

The questionnaires were designed to elicit Soldiers' opinions about their training and experiences with each of the control systems. The questionnaires asked the Soldiers to rate the devices on a 7-point semantic differential scale ranging from "extremely good/easy" to "extremely bad/difficult."

2.3. Procedures

2.3.1. Demographics

Demographic data was taken for each Soldier. Data concerning their physical characteristics and experience, especially their knowledge of operating remote controlled vehicles, was included in the demographic data sheet shown in appendix A.

2.3.2. Control Intuitive Test

Each Soldier was seated and asked to perform the control intuitive test described in section 2.2.2. The Soldier was informed that the purpose of the task was to investigate the words that Soldiers would intuitively use to instruct the robot to perform tasks in specific situations. First, the Soldier was presented a situation and then was asked which words (short and precise instructions) he would use to get the robot to perform the needed task.

2.3.3. Training

No specialized experience was required from the requested Soldiers. Representatives from TAM trained the Soldiers on how to use the robotic simulation, which included both manual and verbal control of the robot. Training questions were included in the post iteration questionnaire so that Soldiers had the opportunity to comment on adequacy of training and provide suggestions for improving the training course.

2.3.4. Speech Control Tasks

After completion of the intuitive test, the Soldiers were asked to perform a series of tasks to assess the contribution of speech control for that task. Tasks were performed using both speech control and manual control. Photographic tasks required the Soldiers to take a picture of an object (an AK-47 and a bomb), label the pictures, enlarge the pictures, and then shrink the pictures. Soldiers were also required to drive the robot to three different waypoints while writing sequential numbers (starting with the number "1") on a sheet of paper. These tasks required simultaneous task accomplishment, the accomplishment of a task while another task was being performed (the secondary task of writing numbers on a sheet of paper while maneuvering the robot), and the activation of menu items found in second and third tiers of the MOCU menu. Tasks were counterbalanced by having Soldiers with odd roster numbers perform the tasks using speech control first and Soldiers with even roster numbers perform the tasks using manual control first. All tasks were timed.

2.3.5. Experimental Design

The experiment was a repeated measure, within subjects design. The independent variable was the type of robotic control (speech or manual). Dependent variables included the following:

- Time to perform each task with and without speech control
- Number and type of initial incorrect control inputs for each task on the initial intuition test
- Number and type of initial incorrect control inputs for each task on the intuition test after training
- Data collector comments
- Questionnaire ratings and comments

3. RESULTS

3.1. Analysis

All objective data collected during the speech control task event were analyzed using paired comparison t-tests. Cohen's d, an index of effect size, was computed for each t-test. Iteration effects were controlled through the counterbalanced order of the experimental design. Soldier questionnaire data were analyzed using descriptive statistics on the subjective ratings.

3.2. Demographics

The participants came from a variety of military occupational specialties and jobs including Infantry, Ranger instructor, pathfinder instructor, team leader, medic, combat engineer, and air assault instructor. The mean number

3.3. Control Intuitiveness Test

Table 1 presents the most frequent phrases suggested by the Soldiers to command the robot in the situations presented. The first column presents the hypothetical situations presented to the Soldiers. The second column presents the current command phrases used to operate the robot in those situations. The phrases suggested by the Soldiers to command the robot in the situations presented are found in the third column and number of Soldiers who suggested each phrase are in the fourth. A complete table containing all the responses can be found in appendix B.

Table 2 shows the nouns and verbs that were most frequently used by the Soldiers compared to the nouns and verbs being used in the current command phrase. A complete table containing all the responses can be found in appendix C.

Table 3 shows the number of Soldiers who responded with the correct command phrase to the hypothetical situations after training and after completion of the timed simulation. A complete listing of all the responses given by the Soldiers after training can be found in appendix D.

The majority of the Soldiers thought that "Take picture" was the most intuitive phrase for situation "a." Once trained, all of the Soldiers but four remembered the current command phrase. Thus, Soldiers and the programmers of the current command phrase were in agreement that the phrase was intuitive and easy to remember.

A little less than half of the Soldiers thought that "Label alpha" was the most intuitive phrase for situation "b." The most frequently suggested verbs, in order, were "label," "name," and "save". While the most frequently used noun was "alpha," many Soldiers used a noun designator before the word "alpha" (i.e., "Label picture alpha," "Label image alpha," etc.). In this situation, the designator is not necessary, because the picture is chosen on the display and the phrase "Label alpha" is more efficient. After being trained, all but one Soldier remembered the current phrase.

Table 1. Most frequent phrases suggested by Soldiers for the hypothetical situations

Hypothetical Situation	Current Command Phrase	Most Frequent Suggested Phrase	No. of Soldiers Suggesting the Phrase
a. You're in a remote location from the robot and see an item that may be of military interest on the screen. You're not sure that you recognize the item and you want to save an image of it to look at it later. What do you tell the robot to do?	Take picture	Take picture	17
		Capture image	3
		Take photo	2
		Take image	2
b. You have a picture of an item on the robot display and you want to name it "A" (Alpha) so you can later identify it to the robot. What do you tell the robot to do?	Label alpha	Label alpha	12
		Save picture, label alpha	2
		Name image alpha	2
c. You want to put the robot in a mode that will allow you to use the mouse to draw an area on the map that is dangerous (IED's are present). By drawing this area on the map, it will keep the robot from entering the area. How do you tell the robot to allow you to draw in this mode?	Activate exclusion zone	Draw mode	4
		Draw	2
		Avoid area	4
d. You have completed the drawing, are satisfied with it, and want the robot to keep out of the area. What do you tell the robot to do?	Execute exclusion zone	Danger area	2
		Stay out of area	2
		Stay clear	2
e. You want to place a line on the map to show that you want the robot to travel along the line from waypoint "A" to waypoint "B." What do you tell the robot to do?	Add route	Draw route	2
		Draw line	2

Hypothetical Situation	Current Command Phrase	Most Frequent Suggested Phrase	No. of Soldiers Suggesting the Phrase
f. The robot is looking around on its own in the autonomous mode and it has generated multiple dots on the map. These dots show you where the robot intends to go. One of the points is in an area where you know IEDs are present and you never want the robot to go in that area. You want the point to be completely removed from the map. What do you tell the robot to do?	Remove goal	Delete point	6
		Remove point	4
		Delete waypoint	2
g. You know generally where enemy activity is the highest and one of the points that the robot generates to let you know where it is going is not as important as the others. You want the robot to go to the other points first but don't want to completely remove the point from the map	Skip a goal	Prioritize point(s)	4
		Disregard point	2
		Autonomous mode	3
		Scan area	3
h. You want the robot to start looking around on its own. What do you tell the robot to do?	Activate self exploration	Scan sector	2
		Roam	2
		Show position	3
i. You are in map view and want to show where the robot is located on the map. What do you tell the robot to do?	Locate the position of the robot on the map	Give location	3
		ID location	2
		Location	2
		Return [to] start	2

Table 1. (Continued)

Hypothetical Situation	Current Command Phrase	Most Frequent Suggested Phrase	No. of Soldiers Suggesting the Phrase
j. You want the robot to come back where it started from. What do you tell the robot to do?	Return home/retrotraverse	point	6
		Return	5
		Return to start	3
		Return to base	3
k. The robot is driving straight ahead. It's going to hit the wall if it keeps going in that direction. You want it to go through the door directly to the right of the robot. What do you tell the robot to do?	Right turn	Turn right	10
		Move right	4
		Go right	3
		Enter door on right	2

Table 2. Parts of speech comparison

Current Command Phrase	Verb	No. of Soldiers Using Word	Noun or adjective/noun	No. of Soldiers Using Word
a. Take picture	take	21	picture	17
	capture	4	image	8
b. Label alpha	label	17	alpha	26
	name	6	picture	6
	save	6	image	5

Current Command Phrase	Verb	No. of Soldiers Using Word	Noun or adjective/noun	No. of Soldiers Using Word
c. Activate exclusion zone	draw	6	draw mode	5
	allow	2	danger area	4
d. Execute exclusion zone	stay	7	area	8
	avoid	4	danger area	4
e. Add route	draw	5	line	8
	travel	4	route	7
	Follow	4	"A", "B", alpha or bravo	6
	input	2	waypoint	4
	move	2	point	2
	execute	2	—	—
f. Remove goal	delete	9	point	14
	remove	6	waypoint	3
	exclude	2	area	2
	avoid	2	danger point	1
g. Skip a goal	prioritize	5	point(s)	12
	go	3	priority	5
h. Activate self exploration	recon	7	Autonomous mode	4
	roam	2	Autonomous search	2
	—	—	sector	2
	—	—	area	3

Table 2. (Continued)

Current Command Phrase	Verb	No. of Soldiers Using Word	Noun or adjective/noun	No. of Soldiers Using Word
i. Locate the position of the robot on the map	give	4	location	9
	id	5	position	6
	display	2	current location	2
	show	4	point	2
	locate	2	—	—
j. Return home/retro traverse	return	18	start point	7
	—	—	start	4
	—	—	Home	3
	—	—	base	3
k. Right turn	turn	12	right	22
	go	4	door[way]	4
	move	4	—	—

Table 3. Phrases used by Soldiers in response
to hypothetical situation after training

Current Command Phrase	Number of Soldiers Using the Phrase
a. Take picture	21
b. Label alpha	24
c. Activate exclusion zone	0
d. Execute exclusion zone	0
e. Add route	0
f. Remove goal	1
g. Skip a goal	2
h. Activate self exploration	10
i. Locate the position of the robot on the map	10
j. Return home/retrotraverse	14
k. Right turn	18

None of the Soldiers responded with the correct phrase during the intuition phase of the experiment and none remembered the current command phrase for situation "c" after being trained. The most common response to the situation during the intuition phase was "Draw mode." This phrase shows that the Soldiers' schemas put the system into a drawing mode similar to the way that Microsoft Windows lets them choose to draw a line or a shape. The command "Draw mode" would allow more flexibility for situation "c" and other situations that require the Soldier to draw or write on the map area.

None of the Soldiers responded with the current command phrase for situation "d" during the intuitive phrase or after being trained. "Exclusion zone" is not a phrase typically used by the military. Soldiers frequently called the off-limits section an area, a danger area, or a danger zone. The most frequent verbs used were avoid, stay out, and stay clear. Designers of the current phrases for situations "c" and "d" were more specific than the Soldiers. They programmed the system to create a specific item (exclusion zone). The Soldiers, on the other hand, thought in terms of telling the system to go into a drawing mode and then telling the system what the drawing meant (how to treat the drawing). The designers then told the system to execute a specific command while the Soldiers told the robot to stay out of an area that was drawn because of what it was labeled. "Avoid danger area" would be a more intuitive command for Soldiers than "Execute exclusion zone." An even more efficient command would be "Draw danger area" for situations "c" and "d." The word "draw" would put the system into the drawing mode and the words "danger area" would command the system to label the drawn area and avoid the area.

Situation "e" was similar to situations "c" and "d." None of the Soldiers used the current command phrase "Add route" during the intuition phase or after being trained. The most common response during the intuition phase was "Draw route." This word "draw" would also put the system into the drawing mode and the word "route" would command the system to label the drawn line as a route and to treat it as such.

The current command phrase for situation "f" was "Remove goal." While none of the Soldiers responded with that phrase, it was primarily because they were not familiar with the concept of intended goals for the robot (areas that the robot intended to explore during autonomous search). The most common response was "Delete point." The verb "delete" was used nine times while the verb "remove" was used six times. Once the Soldiers were trained on the concept of goals, they still did not remember the current command phrase. In fact, only three of them even used the word "goal." The word "point" was still the most commonly used noun, even after training.

Again for situation "g," the Soldiers did not understand the concept of a goal during the intuition phase. None of the Soldiers responded with the current command phrase during the intuition phase and only two responded with the correct phrase after training. The most commonly used noun was the word "point." Soldiers also did not think in terms of "skipping a goal." Instead, they thought more in terms of "prioritizing points" during the intuition phase.

The words the Soldiers used most frequently for situation "h" did not adequately describe what the robot needed to do. The most popular verb used was "scan" and this verb is more commonly associated with panning a camera rather than driving around to explore an area. They also used unexplained nouns (e.g., area, sector, and perimeter) as adverbs. The most frequently used noun was "autonomous" and this was not an adequate word, since the robot was already in the autonomous mode. The current command phrase (activate self exploration) was not chosen initially, but after training, 10 of the Soldiers used the correct command phrase. During the intuition phase, a few of the Soldiers suggested that the words "recon" or "recon mode" be used and this would probably be the most descriptive phrase and the one that would be closest to military phraseology.

The current command phrase for situation "i" (Locate the position of the robot on the map) is quite wordy. The Soldiers' suggestions during the intuition phase were much terser. The word "position" was the most frequently used noun and the verbs "give," "ID," and "show" were the most frequently used verbs. The phrase "show position" would be adequate if the display was in map view. The current command phrase was remembered by 10 of the Soldiers after training.

While only one Soldier suggested using the word "ping," it would be the most descriptive and efficient way to get the idea across. The word "ping" is used in the naval world, the gaming world, and the computer world. In the naval and gaming lexicon, "ping" comes from a submarine sonar search. A short sound burst is sent and an echoing ping returns so the submariner knows where the object pinged is located (range to the target). In the computer world, "ping" is a program that allows a user to verify that another computer is reachable by sending it a message and waiting for an acknowledgment. Before this word is chosen, research would be needed to determine if the Soldier target audience readily understands its meaning.

For situation "j" the current command phrase was "Return home/retrotraverse." Two Soldiers suggested using just "Return home;" none used the word retrotraverse during the intuition phase. The verb "return" was used by 17 of the Soldiers in the intuition phase, while the most frequently used noun was "start/start point." Fourteen of the Soldiers remembered "Return home/retrotraverse" after training.

For situation "k" the current command phrase was "Right turn." Ten of the Soldiers suggested "Turn right" and almost all of the Soldiers put their suggested verb in front of the direction. Even after being trained on the current phrase, seven of the Soldiers still placed the verb before the noun. In fact, during the entire intuition phase, the vast majority of the Soldiers placed the verb first in their suggested phrase for every command in each of the situations.

3.4. Training

Soldiers felt that the speech-based control training was very good. Their ratings are shown in table 4.

One of the 29 Soldiers participating in the evaluation indicated that he needed a little more practice. Another rated the training as neutral, because he had difficulty pronouncing the words (English was his second language). Three Soldiers indicated that the manual controller was the hardest training task to learn and two indicated that the speech control was the hardest task.

Twenty-eight out of the 29 Soldiers who trained on the speech-control system were easily understood by it. (One Soldier's voice had a compatibility issue with the microphone being used and did not participate in the evaluation. That issue has since been resolved.) Forty sentences were used to train the system to the Soldiers' voices and this was accomplished in 7 to 10 min.

Table 4. Soldiers' ratings of the training

Category	Mean Rating
Completeness of speech training	5.78
Comprehension of the concept of the speech control	5.88
Overall evaluation of the speech training course	5.81

3.5. Speech Control Tasks

The average times to complete tasks using verbal and manual control are shown in table 5. Table 6 is a summary of the paired-sample t-tests used to compare the mean verbal and manual values for each task. Holm's Bonferroni procedure was used to control for family-wise error. Cohen's d is an index of effect size, the difference in the two means divided by the pooled SD. For the photographic tasks, the verbal control was faster than the manual control for all tasks except taking the photograph (this task required moving the robot into position in order to take the picture). The manual control was significantly faster for taking the photograph. There was no significant difference between the times required to drive to the waypoints.

Table 5. Mean times to complete tasks

Task	Verbal Mean (s)	SD	Manual Mean (s)	SD
Take a picture	15.5	3.1	13.4	3.2
Label the picture	3.8	1.4	7.7	3.3
Enlarge the picture	2.8	1.7	5.7	1.7
Shrink the picture	2.1	0.9	6.2	2.6
Drive to two waypoints	133.6	47.3	119.4	21.0

Intuitive Speech-Based Robotic Control

Table 6. Summary of paired-sample *t*-tests

Task	t	df	obtained p	required p	d
Take a picture	2.87	28	0.008a	0.025	0.65
Label the picture	−5.23	28	< 0.001a	0.0167	1.54
Enlarge the picture	−7.74	28	< 0.001a	0.0125	1.78
Shrink the picture	−7.83	28	< 0.001a	0.010	2.11
Drive to two waypoints	1.60	28	0.12	0.050	0.39

[a] $p < 0.05$, 2-tailed.

Table 7 summarizes the secondary task (the mean number of numbers written on the paper while driving to the waypoints). Performance on the secondary task (writing down numbers) while driving was significantly better with the verbal control. Table 8 is a summary of the paired-sample *t*-test used to compare the mean verbal and manual values for the secondary task.

Table 7. Mean numbers written per second while driving

	Verbal		Manual	
	Mean	SD	Mean	SD
Numbers per second	0.47	0.14	0.39	0.13

Table 8. Summary of paired-sample *t*-test

Task	t	df	p	d
Numbers per second	3.671	28	0.001a	0.59

[a] $p < 0.05$, 2-tailed.

4. DISCUSSION AND RECOMMENDATIONS

This experiment demonstrated how important it is to tailor speech commands to the target audience. Before training, less than 10% of the commands the Soldiers thought should be used were the commands that were programmed into the speech-control system. Even after training and using many of the commands during a simulation task, only 34% of the Soldiers remembered the commands that the system designers programmed.

Commands that were initially intuitive ("Take picture" and "Label alpha") were correctly used by 72% and 83%, respectively, of the Soldiers after training. Conversely, less intuitive phrases such as "Activate exclusion zone" were not remembered by any of the Soldiers, even after training. Thus, we determined that it is important to develop intuitive command phrases that are based on military phrases and the Soldiers' schemas, because doing so would result in fewer errors and reduce the time it takes Soldiers to perform robotic tasks, especially during times of combat stress and high cognitive load.

The use of grammatical rules or language models to organize the command words would help make them more meaningful and easier for the Soldiers to remember. For example, interestingly, the vast majority of the Soldiers placed the verb first in their suggested command phrases. A simple rule that states "always place the verb before the noun" would be a good rule to institute, because it would provide consistency and follow what the Soldiers tend to do naturally. The verb tells the robot what behavior it should perform and the following word tells the robot where or how it should perform the behavior. "Turn" tells the robot that it will change directions and "right" tells the robot which direction. "Draw" tells the robot to go into the drawing mode and "danger area" tells the robot what the resulting drawing is and how to treat the drawing. "Forward" tells the robot to move in a forward direction and "10 ft" tells the robot how far to go forward. "Go" tells the robot to turn on the driving behavior and "home" tells the robot where it is suppose to drive. These "rules" should be as consistent as possible with the Soldiers' natural language and be consistent between commands.

For the photographic tasks, the verbal control was faster than the manual control for all tasks except taking the photograph. The manual control was significantly faster for taking the photograph, because this task required driving the robot into position in order to take the picture. This is consistent with the literature, because driving has been shown to be more efficient with manual control than with speech-control (Pettitt et al., 2009). A continuous task such as turning requires either starting a behavior that continues until stopped (having to stop the second the robot gets into the correct position with no lag) or repeating a command multiple times (if the robot only turns a few degrees each time it is told to turn). The mean driving time was larger for speech control than for manual control, but there was no significant difference between the total times, partially because of the large variance in driving times using speech control. However, when the driving times for each single waypoint were compared, the manual control was significantly faster for driving to the blue waypoint, but this result was not true for the green waypoint. This difference might be explained by the fact that driving to the blue waypoint required greater maneuvering around an obstacle that was located directly in front of the unmanned ground vehicle (UGV). This setup forced the Soldiers to do a lot of turning and negotiating. Driving to the green waypoint was primarily a straight route that only required slight turns around the obstacle.

Overall, all three of our hypotheses were met:

1) Soldiers were able to perform a secondary task (writing numbers) significantly faster when operating the robot using speech control than they were when operating the robot using manual control. It is clear that robotic control requires multitasking. It also appears that speech control required less attention than manual control, thus freeing up cognitive resources for additional tasks.

2) Speech control allowed significantly faster task performance than manual control when the task involved the use of menu items (enlarge picture, shrink picture). Speech control allowed direct access to the menu items, while manual control required navigating through a menu and selecting an item that was two levels deep into the menu. Speech control was also significantly faster for labeling items in which Soldiers had to choose a list and then select Alpha, Bravo, Charlie, Delta, or Echo from the list to label the picture.

3) Speech control took significantly longer when performing continuous tasks such as turning the robot during the "take a picture" task and driving to the blue waypoint, which involved a significant amount of turning.

When interpreting the results, it is important to consider the tasks that were examined during this experiment. The tasks used in the intuition and speech-control portions of the experiment were tasks that could be found in a robotic reconnaissance mission and the findings are specific to these tasks.

5. CONCLUSIONS

- It is extremely important to tailor speech commands to the target audience. Tailoring allows better retention and more efficient operation.

- Speech control is quicker than manual control in situations that require secondary task accomplishment and also in situations in which the items that need to be accessed are embedded in menus.
- Manual control is more effective than speech control for non-discrete tasks such as turning.

DISCLAIMERS

The findings in this report are not to be construed as an official Department of the Army position unless so designated by other authorized documents.

Citation of manufacturer's or trade names does not constitute an official endorsement or approval of the use thereof.

REFERENCES

Apaydin, O. *Networked Humanoid Animation Driven By Human Voice Using Extensible 3D(X3D), H-Anim and Java Speech Open Standards*; ADA401793; Naval Postgraduate School: Monterey, 2002.

Bortolussi, M. R.; Vidulich, M. A. The Effects of Speech Controls on Performance in Advanced Helicopters in a Double Stimulation Paradigm. *International Symposium on Aviation Psychology*, Columbus, OH, 29 April–2 May 1991, 6, 216–221.

Bourakov, E.; Bordetsky, A. Voice-on-Target: A New Approach to Tactical Networking and Unmanned Systems Control Via the Voice Interface to the SA Environment. *Proceedings of the 14th International Command and Control Research and Technology Symposium (ICCRTS)*, Washington, DC, 15–17 June 2009.

Cassenti, D.; Kelley, T.; Swoboda, J.; Patton, D. Submitted to *Proceedings of 53rd Annual Meeting of the Human Factors and Ergonomics Society*. Santa Monica, CA: Human Factors and Ergonomics Society, October 19–23, 2009.

Chen, Y. C.; Hass, E. C.; Pillalamarri, K.; Jacobson, C. N. *Human-Robot Interface: Issues in Operator Performance, Interface Design, and Technologies;* ARL-TR-3834; U.S. Army Research Laboratory: Aberdeen Proving Ground, MD, 2006.

Graham, R.; Carter, C. Comparison of Speech Input and Manual Control of In-Car Devices While on the Move. *Personal and Ubiquitous Computing* 2006, 4, 155–164.

Haas, E. C.; Edworthy, J. *The Ergonomics of Sound: Selections from Human Factors and Ergonomics Society Annual Meetings.* Santa Monica, CA, 1985–2000, 2002.

Henry, P. P.; Mermagen, T. J.; Letowski, T. R. *An Evaluation of a Spoken Language Interface;* ARL-TR-3477; U.S. Army Research Laboratory: Aberdeen Proving Ground, MD, 2005.

Karis, D.; Dobroth, K. M. Psychological and Human Factors Issues in the Design of Speech Recognition Systems; In *Applied Speech Technology;* Syrdal, A.; Bennett, R.; Greenspan, S. Eds.; CRC Press: Ann Arbor, MI; 1995, pp. 359–388.

Myers, G. K.; Cowan, C. *Robotic Collaborative Technical Alliance Quarterly Technical Status Report;* 8752-079-43-25; Stanford Research Institute International: Menlo Park, CA, 1 July– 30 September 2003.

Nielsen, J.; Sano, D. SunWeb: User Interface Design for Sun Microsystems' Internal Web. *Proceedings of the Second World-Wide Web Conference,* 1994.

Noyes, J. M.; Baber, C.; Leggatt, A. P. Automatic Speech Recognition, Noise and Workload. *Proceedings of the 44th Annual Meeting of the Human Factors and Ergonomics Society.* Santa Monica, CA, 2000, pp 3762–3765.

Pettitt, R. A.; Redden, E. S.; Carstens, C. B. *Scalability of Robotic Controllers: Speech-Based Robotic Controller Evaluation*; ARL-TR-4858; U.S. Army Research Laboratory: Aberdeen Proving Ground, MD, 2009.

Sams, R. *Ground Robotics Technology Demonstration.* Limited User Assessment; U.S. Marine Corps Forces: Pacific Experimentation Center, Comap Smith, HI, 2009.

Steeneken, H. J. M. *Potentials of Speech and Language Technology Systems for Military Use: An Application and Technology Oriented Survey;* AC/243 (Panel 3) TR/21; North Atlantic Treaty Organization Defence Research Group Panel 3 on Physics and Electronics: The Netherlands, 1996.

Vloeberghs, C.; Verlinde, P.; Swail, C.; Steeneken, H.; South, A. The Impact of Speech Under "Stress" on Military Speech Technology; *RTO-TR-10* AC/323(IST)TP/5; NATO Research and Technology Organization: The Netherlands, 2000.

APPENDIX A.
DEMOGRAPHICS QUESTIONNAIRE[*]

Sample Size = 29

	MOS	RANK	DUTY POSITION	
09S – 10	68D – 1	E4 – 6	Air Aslt Instructor – 1	Pathfinder Instructor - 1
11B – 8	68W – 1	E5 – 17	Crew Chief – 1	Ranger Instructor – 1
11C – 1	88M – 2	E6 – 2	Driver – 1	Support – 1
12B – 1	0CS – 3	OCS – 4	Instructor – 1	Team leader – 4
15T – 1			Medic – 1	Training – 1
31B – 1			OCS – 7	NR - 8
			Combat Engr - 1	

AGE

28 years (range 21-47)

1) How long have you served in the military? 65 months (mean)
2) How long have you had an infantry-related job? 72 months (mean)
3) How long have you been a fire team leader? 21 months (mean)
4) How long have you been a squad leader? 44 months (mean)
5) How long have you been deployed overseas? 21 months (mean)
6) How long have you been deployed in a combat area? 15 months (mean)
7) With which hand do you most often write? 24 Right 5 Left
8) With which hand do you most often fire a weapon? 26 Right 3 Left
9) Do you wear prescription lenses? 4 Yes 25 No
10) If yes, which do you wear most often? 2 Glasses 2 Contacts
11) Which is your dominant eye? 25 Right 4 Left
12) Do you have any vision related problem? 2 Yes 27 No If so, what? Red/green color blind (1), farsighted (1)
13) Have you ever used a robotic system? 1 Yes 23 No 5 NR If so, what type? Davinci (1)

---

[*] This appendix appears in its original form without editorial change.

14) Please describe the conditions under which you used the robotic system. Surgical (1)
15) Using the scale below, please rate your skill level for each of the following activities.

None	Beginner	Intermediate	Expert
1	2	3	4

ACTIVITY	MEAN RESPONSE
Operating ground unmanned vehicles	1.13
Operating aerial vehicles	1.09
Target detection and identification	1.55
Playing commercial video games	2.65
Training with Army video simulations	2.17

APPENDIX B. PHRASES SUGGESTED BY SOLDIERS FOR THE HYPOTHETICAL SITUATIONS[*]

(Note that not all responses add to 29 as sometimes Soldiers responded more than once and others times some Soldiers did not respond.)

Hypothetical Situation	Current Command Phrase	Soldier's Suggested Command Phrase	Number of Soldiers Suggesting the Phrase
a. You're in a remote location from the robot and see an item that may be of military interest on the screen. You're not sure that you recognize the item and you want to save an image of it to look at it later. What do you tell the robot to do?	Take picture	Take picture	17
		Take photo	2
		Take image	2
		Image	1
		Capture image	3
		Capture	1
		Save image	2

---

[*] This appendix appears in its original form without editorial change.

Appendix B. (Continued)

Hypothetical Situation	Current Command Phrase	Soldier's Suggested Command Phrase	Number of Soldiers Suggesting the Phrase
b. You have a picture of an item on the robot display and you want to name it "A" (Alpha) so you can later identify it to the robot. What do you tell the robot to do?	Label alpha	Label alpha	12
		Name alpha	1
		Save alpha	1
		Save picture, label alpha	2
		Save picture alpha	1
		Label off	1
		Label image	1
		Label image alpha	1
		Save as image alpha	1
		Save as alpha	1
		Label picture alpha	1
		Name image alpha	2
		Name picture alpha	1
		Save photo as alpha	1
		Name alpha	1
		Name picture	1
c. You want to put the robot in a mode that will allow you to use the mouse to draw an area on the map that is dangerous (IED's are present). By drawing this area on the map, it will keep the robot from entering the area. How do you tell the robot to allow you to draw in this mode?	Activate exclusion zone	Draw mode	4
		Allow draw mode	1
		Draw	2
		Draw map	1
		Draw line	1
		Draw picture	1
		Observe and draw	1

		Manual draw	1
		Manual mode	1
		Allow mouse	1
		Mouse control	1
		Use mouse	1
		Mouse mode	1
		Mouse	1
		Danger zone	1
		Danger area	1
		Label danger area	1
		Enter danger area	1
		Danger area drop	1
		Restricted mode	1
		Quarantine mode	1
		Map restrictions	1
		Shade area on map	1
		Proceed avoiding dangerous map display	1
		Flag area	1
		Standby	1
d. You have completed the drawing, are satisfied with it, and want the robot to keep out of the area. What do you tell the robot to do?	Execute exclusion zone	Avoid area	4
		Restricted area	1
		Restrict area	1
		Danger area	2
		Exit danger area	1
		Keep out of danger area	1
		Danger area lock	1
		Designate area	1
		Steer clear	1
		Stay out of area	2

Appendix B. (Continued)

Hypothetical Situation	Current Command Phrase	Soldier's Suggested Command Phrase	Number of Soldiers Suggesting the Phrase
		Stay	1
		Stay out of shaded area	1
		Stay out of selected area	1
		Stay clear	2
		Keep out of area	1
		Keep out of exclusion zone	1
		Do not enter map	1
		Follow last map restrictions	1
		Send drawing	1
		Follow instructions	1
		Area forbidden	1
		Seek hull	1
e. You want to place a line on the map to show that you want the robot to travel along the line from waypoint "A" to waypoint "B." What do you tell the robot to do?	Add route	Draw route	3
		Draw line	2
		Draw waypoint	1
		Input path	1
		Input line	1
		Line	1
		Mouse control	1
		Mark point	1
		Yellow route designated	1
		Drive route	1
		Follow route	1
		Follow line	1
		Travel line	1

Intuitive Speech-Based Robotic Control

		Travel on dark	1
		Move from point "A" to point "B"	1
		Travel from "A" to "B"	1
		Travel "A" to	1
		Traverse from route "A" to "B"	1
		Follow waypoint to waypoint	1
		Go from waypoint alpha to bravo	1
		Move to waypoint	1
		Execute last trace	1
		Execute	1
		Follow directions	1
		Standby	1
		Scroll to line	1
f. The robot is looking around on its own in the autonomous mode and it has generated multiple dots on the map. These dots show you where the robot intends to go. One of the points is in an area where you know IEDs are present and you never want the robot to go in that area. You want the point to be completely removed from the map. What do you tell the robot to do?	Remove goal	Delete point	6
		Delete waypoint	2
		Remove point	4
		Remove	1
		Remove exclusion point	1
		Exclude point	1
		Avoid point	1
		Avoid dot on	1
		Modify route	1
		Delete area	1
		Do not enter	1
		Eliminate points	1
		Input path	1
		Filter last overlay	1
		Cancel	1
		Mark restricted	1
		Restrict area	1

Appendix B. (Continued)

Hypothetical Situation	Current Command Phrase	Soldier's Suggested Command Phrase	Number of Soldiers Suggesting the Phrase
g. You know generally where enemy activity is the highest and one of the points that the robot generates to let you know where it is going is not as important as the others. You want the robot to go to the other points first but don't want to completely remove the point from the map. What do you tell the robot to do?	Skip a goal	Prioritize point(s)	4
		Prioritize on my command	1
		Priority low	1
		Follow by	1
		Priority area	1
		Low priority	1
		Reorganize priority	1
		Set priority	1
		Priority point	1
		Disregard point	2
		Sequence points	1
		Go to high activity points	1
		Go to another point	1
		Remove point	1
		Hold point	1
		Skip point	1
		Stop and mark waypoint	1
		Stop and redirect	1
		Redirect	1
		Follow by interest level	1
		Go around	1
		Detour	1
		Reverse order	1
		Do not enter	1

Intuitive Speech-Based Robotic Control

Hypothetical Situation	Current Command Phrase	Soldier's Suggested Command Phrase	Number of Soldiers Suggesting the Phrase
h. You want the robot to start looking around on its own. What do you tell the robot to do?	Activate self exploration	Roam	2
		Autonomous mode	3
		Perform autonomous mode	1
		Autonomous search mode	1
		Autonomous search	1
		Enter autonomously	1
		Scan area	3
		Scan sector	2
		Scan around	1
		Begin scan	1
		Scan	1
		Area recon	1
		Recon mode	1
		Recon	1
		Self mode	1
		Auto pilot	1
		Set up patrol of perimeter area	1
		Search perimeter	1
		Explore	1
		Unassisted	1
		Go free	1
		Site survey	1

Appendix B. (Continued)

Hypothetical Situation	Current Command Phrase	Soldier's Suggested Command Phrase	Number of Soldiers Suggesting the Phrase
i. You are in map view and want to show where the robot is located on the map. What do you tell the robot to do?	Locate the position of the robot on the map	Ping	1
		Display current location	1
		Location	2
		Location point	1
		Transmit location	1
		Plot location	1
		Give location	3
		Current location	1
		ID location	2
		ID self	1
		ID robot	1
		ID on map	1
		ID locate	1
		Locate point	1
		Drop point	1
		Show position	3
		Show current position	1
		Mark your position	1
		Give position	1
		Display position	1
		Map view	1
j. You want the robot to come back where it started from. What do you tell the robot to do?	Return home/retrotraverse	Return to start	3
		Return [to] start point	6
		Return to base	3
		Return to loading point	1
		Return home	2
		Home	1
		Return	5

Intuitive Speech-Based Robotic Control

		Go back to start point	1
		Go to starting point	1
		Restart	1
		Starting	1
		Alignment start point	1
		Locate origin	1
		Reset	1
		Reverse	1
k. The robot is driving straight ahead. It's going to hit the wall if it keeps going in that direction. You want it to go through the door directly to the right of the robot. What do you tell the robot to do?	Right turn	Turn right	10
		Turn	1
		Go right	3
		Move right	4
		Take [a] right	1
		Right door	1
		Reroute direction	1
		Redirect through doorway	1
		Enter door on right	2
		Turn 45 degrees	1
		Right 90 degrees	1
		Right go	1

APPENDIX C. PARTS OF SPEECH COMPARISON[*]

Current Command Phrase	Verb	No. of Soldiers Using Word	Noun or adjective/noun	No. of Soldiers Using Word
a. Take picture	take	21	picture	17
	capture	4	image	8
	save	2	photo	2
b. Label alpha	label	17	alpha	26
	name	6	picture	6
	save	6	image	5
			photo	1

---

[*] This appendix appears in its original form without editorial change.

Appendix C. (Continued)

Current Command Phrase	Verb	No. of Soldiers Using Word	Noun or adjective/noun	No. of Soldiers Using Word
c. Activate exclusion zone	draw	6	draw mode	5
	label	1	manual mode	1
	allow	2	restricted mode	1
	observe	1	quarantine mode	1
	use	1	mouse mode	1
	enter	1	mouse control	1
	drop	1	mouse	2
	shade	1	danger area	4
	proceed	1	danger zone	1
	flag	1	map restrictions	1
	standby	1	area	2
			dangerous map display	1
d. Execute exclusion zone	avoid	4	area	8
	stay	7	danger area	4
	restrict	1	restricted area	1
	designate	1	Shaded area	1
	exit	1	Selected area	1
	keep	2	map	1
	steer	1	map restrictions	1
	enter	1	drawing	1
	follow	1	instructions	1
	send	1	hull defilade	1
	seek	1		
e. Add route	draw	5	line	8
	travel	4	route	7
	follow	4	"A", "B", alpha or bravo	6
	input	2	waypoint	4
	move	2	point	2
	execute	2	path	1
	scroll	1	mouse	1
	drive	1	trace	1
	go	1	exclusions	1

Intuitive Speech-Based Robotic Control

	traverse	1		
	stand	1		
f. Remove goal	delete	9	point	14
	remove	6	waypoint	3
	exclude	2	exclusion point	1
	avoid	2	danger point	1
	eliminate	1	area	2
	modify	1	restricted area	1
	enter	1	route	1
	input	1	direction	1
	filter	1	overlay	1
	cancel	1	movement	1
	mark	1	path	1
	restrict	1	dot	1
g. Skip a goal	prioritize	5	point(s)	12
	go	3	priority	5
	stop	2	priority area	1
	follow	2	priority point	1
	disregard	2	waypoint	1
	sequence	2	command	1
	redirect	2	order	1
	hold	1	interest level	1
	mark	1		
	reorganize	1		
	remove	1		
	detour	1		
	enter	1		
	reverse	1		
	set	1		
	skip	1		
h. Activate self exploration	scan	1	autonomous mode	4
	recon	7	autonomous search	2
	roam	2	auto pilot	1
	enter	1	area recon	1
	explore	1	recon mode	1
	set	1	self mode	1

Appendix C. (Continued)

Current Command Phrase	Verb	No. of Soldiers Using Word	Noun or adjective/noun	No. of Soldiers Using Word
	search	1	sector	2
	perform	1	area	3
	go	1	perimeter	1
	survey		site	1
			perimeter area	1
			patrol	1
i. Locate the position of the robot on the map	give	4	location	9
	id	5	location point	1
	display	2	current location	2
	transmit	1	current position	1
	plot	1	position	6
	show	4	point	2
	drop	1	map	1
	mark	1	robot	1
	locate	2	self	1
			map view	1
j. return home/retrotraverse	return	18	home	3
	go back	1	start point	7
	restart	1	starting point	1
	starting	1	start	4
	alignment	1	loading point	1
	locate	1	base	3
	reset	1	origin	1
k. right turn	turn	12	right	22
	go	4	direction	1
	move	4	45 degrees	1
	take	1	90 degrees	1
	enter	2	door[way]	4
	reroute	1		
	redirect	1		

APPENDIX D. PHRASES USED BY SOLDIERS IN RESPONSE TO HYPOTHETICAL SITUATIONS AFTER TRAINING[*]

Hypothetical Situation	Current Command Phrase	Soldier's Command Phrase After Training	Number of Soldiers Suggesting the Phrase
a. You're in a remote location from the robot and see an item that may be of military interest on the screen. You're not sure that you recognize the item and you want to save an image of it to look at it later. What do you tell the robot to do?	Take picture	Take picture	21
		Label alpha	1
		Save picture	2
		Take image	1
b. You have a picture of an item on the robot display and you want to name it "A" (Alpha) so you can later identify it to the robot. What do you tell the robot to do?	Label alpha	Label alpha	24
		Save alpha	1
c. You want to put the robot in a mode that will allow you to use the mouse to draw an area on the map that is dangerous (IED's are present). By drawing this area on the map, it will keep the robot from entering the area. How do you tell the robot to allow you to draw in this mode?	Activate exclusion zone	Something mode	1
		Draw	1
		Draw area	1
		Mark area	1
		Exclusion out	1
		Robot control	1
		No response	19
d. You have completed the drawing, are satisfied with it, and want the robot to keep out of the area. What do you tell the robot to do?	Execute exclusion zone	Keep in boundary	1
		Execute	2
		Avoid area	2
		Do not enter	1
		Exit area	1

---

[*] This appendix appears in its original form without editorial change.

Appendix D. (Continued)

Hypothetical Situation	Current Command Phrase	Soldier's Command Phrase After Training	Number of Soldiers Suggesting the Phrase
		Exclude area	1
		Follow route	1
		Exclusion	1
		Go around exclusion point	1
		Keep out	1
		Exclude point	1
		Keep out mode	1
		Skip area	1
		No response	10
e. You want to place a line on the map to show that you want the robot to travel along the line from waypoint "A" to waypoint "B." What do you tell the robot to do?	Add route	Go to route	1
		Draw line	2
		Draw waypoint	2
		Execute route	1
		Travel to "A" point "A"	1
		Follow points/add points	1
		Follow route	2
		Go to point	1
		Remove execution point	1
		Drive forward	1
		No response	12

Intuitive Speech-Based Robotic Control

Hypothetical Situation	Current Command Phrase	Soldier's Command Phrase After Training	Number of Soldiers Suggesting the Phrase
f. The robot is looking around on its own in the autonomous mode and it has generated multiple dots on the map. These dots show you where the robot intends to go. One of the points is in an area where you know IEDs are present and you never want the robot to go in that area. You want the point to be completely removed from the map. What do you tell the robot to do?	Remove goal	Remove goal	1
		Delete point	3
		Remove point	5
		Delete waypoint	3
		Remove	1
		Eliminate point	1
		Stay out of execution area	1
		Remove execution point	1
		Remove exclusion point	1
		Lead point	1
		Cancel goal	1
		Danger	1
		Delete goal	1
		No response	4
g. You know generally where enemy activity is the highest and one of the points that the robot generates to let you know where it is going is not as important as the others. You want the robot to go to the other points first but don't want to completely remove the point from the map. What do you tell the robot to do?	Skip a goal	Skip a goal	2
		Do not enter	1
		Ignore point	1
		Skip point	5
		Avoid	1
		Go around exclusion point	1
		Avoid point	1
		Ignore goal	1
		Reverse point	1
		Low priority	1
		No response	9

Appendix D. (Continued)

Hypothetical Situation	Current Command Phrase	Soldier's Command Phrase After Training	Number of Soldiers Suggesting the Phrase
h. You want the robot to start looking around on its own. What do you tell the robot to do?	Activate self exploration	Activate self exploration	10
		Robot search	1
		Search around	1
		Search area	2
		Explore	1
		Scan mode	1
		Auto pilot	1
		No response	7
i. You are in map view and want to show where the robot is located on the map. What do you tell the robot to do?	Locate the position of the robot on the map	Locate the position of the robot on the map	10
		Find robot	1
		Robot location	2
		Route robot	1
		Show point	1
		Locate	1
		Location	1
		ID mode	1
		Give position	1
		Show location	1
		Mark position	1
		No response	4
j. You want the robot to come back where it started from. What do you tell the robot to do?	Return home/ retrotraverse	Return home/retrotraverse	14
		Return home	3
		Return to start point	2
		Come home	2
		Come back	1
		Return	1
		No response	2

Hypothetical Situation	Current Command Phrase	Soldier's Command Phrase After Training	Number of Soldiers Suggesting the Phrase
k. The robot is driving straight ahead. It's going to hit the wall if it keeps going in that direction. You want it to go through the door directly to the right of the robot. What do you tell the robot to do?	Right turn	Right turn	18
		Turn right	5
		Go right	1
		Stop-turn right	1

LIST OF SYMBOLS, ABBREVIATIONS, AND ACRONYMS

ARL	U.S. Army Research Laboratory
ASR	automated speech recognition
COTS	commercial-off-the-shelf
HRED	Human Research and Engineering Directorate
MOCU	Multi-Robot Operator Control Unit
OCS	Officer Candidate School
SCR	speech command recognition
TAM	Think-A-Move, Ltd.

INDEX

A

access, 4, 17, 18, 20, 21, 29, 84, 86, 103
algorithm, 25
ANOVA, 44, 45, 46, 48, 51, 82
architecture design, 22, 23
Army Research Laboratory (ARL), vii, 1, 34
assault, 34, 90
assessment, 39, 73, 82
assets, 11, 13, 59, 60, 61, 62
automation, viii, 21, 31, 33, 34, 44, 54
autonomy, vii, 1, 2, 31, 33, 35, 36, 40, 41, 42, 43, 44, 45, 51, 52, 53, 54, 65, 68, 69, 71, 73, 84
avoidance, 35, 36, 42, 53, 76, 77, 79
awareness, 33, 34, 43, 65, 66, 69, 82

B

base, vii, viii, 5, 14, 15, 16, 18, 21, 83, 93, 96, 99, 115, 119
batteries, 2, 11, 60
brain, 14, 15, 16, 18, 22

C

C++, 24
chain of command, 62

chemical, 38, 39, 59, 60
clone, 21
cognitive load, 33, 43, 54, 102
collateral, 33
color, 9, 10, 52, 63, 71, 73, 106
commercial-off-the-shelf (COTS), 87
compatibility, 100
complexity, 24
computational capacity, 3
computer, 4, 5, 11, 12, 16, 28, 37, 39, 70, 85, 87, 99
computing, vii, 1, 2, 4, 9, 25
concurrency, 23, 25
configuration, 16, 17, 18, 21
construction, 18, 21
control condition, 46, 48
coordination, 84
correlation, 47
cost, 6
CPU, 4, 11, 14

D

danger, 94, 95, 97, 102, 109, 116, 117
data structure, 5
decoupling, 24
delegates, 32
Delta, 103
demographic data, 40, 89
depth, 25

derivatives, 10
designers, 32, 97, 101
detection, 2, 7, 28, 53, 54, 64, 107
distribution, 11, 15, 21, 23, 24, 25, 29
drawing, 48, 67, 92, 97, 98, 102, 108, 109, 110, 117, 120

E

echoing, 99
election, 5
e-mail, 21
emergency, 13
environment, vii, 1, 2, 5, 6, 7, 8, 17, 18, 19, 20, 21, 32, 54, 85
exclusion, 92, 94, 97, 102, 108, 109, 110, 111, 116, 117, 120, 121, 122
execution, 121, 122
exercise, 67
experimental design, 44, 90

F

fear, 33, 85
fires, 58, 59, 60
flexibility, 20, 29, 97
flight, 7, 8, 56
fluid, 24
force, 60, 61, 85
foundations, 24, 25
friction, 28
fusion, 10

G

geometry, 5, 6, 7, 8, 9
Georgia, 14
GPS, 2, 10, 32, 35, 53, 82
guidelines, 22, 24

H

hazards, 34

height, 7, 78
history, 20
host, 9, 18, 19, 21, 24
human, viii, 5, 27, 28, 31, 32, 33, 39, 86
Human Research and Engineering Directorate (HRED), viii, 34, 83
human will, 27
hypothesis, 33

I

ideal, viii, 11, 31
identification, 54, 60, 64, 66, 77, 84, 107
image, 5, 8, 9, 15, 35, 78, 87, 88, 91, 92, 94, 107, 108, 116, 120
imagery, 5, 6
independent variable, 90
infrastructure, 22, 23, 25
institutions, 14, 25
integration, 5, 10, 25
intelligence, vii, 1, 22, 23, 61
interface, 4, 5, 16, 28, 35, 36, 37, 42, 77, 87, 88
intervention, viii, 31, 32, 33
issues, 4, 17
iteration, 40, 43, 47, 89

J

Java, 24, 104

L

labeling, 103
laptop, 4, 20, 37
lasers, 32
latency, 33, 52, 53
learning, 34, 67
LED, 78
lens, 5, 6, 9, 77, 78
light, 8, 58, 70, 76, 77
localization, 5, 35
logging, 4, 18
LTC, 62

M

mapping, 5, 7, 35, 36, 42, 45, 47, 48, 53, 54, 65, 74, 81
Marine Corps, 105
mental model, 54
methodology, 22
Microsoft, 37, 42, 87, 97
military, 25, 32, 34, 37, 40, 63, 82, 85, 86, 90, 92, 97, 98, 102, 106, 107, 120
mission, vii, viii, 2, 5, 11,18, 33, 34, 36, 40, 41, 43, 59, 60, 62, 67, 71, 76, 80, 83, 85, 86, 103
mobile device, 84
mobile robots, 7, 16, 25
modules, 2, 16, 22, 25
motor control, 5, 11, 12, 16
Multi-Robot Operator Control Unit (MOCU), 35, 87

N

NATO, 105
negotiating, 103
Netherlands, 105
neutral, 76, 99
North Atlantic Treaty Organization, 105

O

obstacles, 7, 28, 32, 36, 52, 59, 60, 61, 66, 68, 73, 77
OH, 55, 104
operating system, 2, 14, 15, 16
operations, 12, 37, 40, 41, 43, 52, 60, 82
optimal performance, 33
overlay, 112, 117
oxygen, 85

P

Pacific, 34, 35, 41, 55, 82, 87, 105
parallel, 58

parallelism, 24
participants, 34, 43, 45, 47, 48, 54, 90
physical activity, 87
physical characteristics, 40, 89
platform, 2, 6, 16, 18, 21, 22, 25, 32
playing, 72
pop-up windows, 77
Portugal, 26
project, 5, 14, 15, 19, 20, 24, 25

Q

query, 85
question mark, 77
questionnaire, 40, 43, 44, 45, 47, 89, 90

R

radar, 2
reaction time, 33
real time, 7
recognition, 84, 85, 86, 87, 124
reliability, 4, 47, 75
requirements, 17, 18, 19, 23, 27
research institutions, 25
resolution, 5, 6, 7, 8, 9, 52, 70, 72, 76
response, 27, 28, 53, 74, 97, 98, 120, 121, 122, 123
response time, 28
restrictions, 28, 109, 110, 116, 117
robotics, vii, 1, 2, 7, 14, 16, 25, 26
robotics labs, vii, 1
rules, 102

S

safety, 12, 13, 25, 28
schema, 32
scripts, 15, 19, 20, 21
Second World, 105
Secretary of Defense, 55
security, 21, 33, 59, 61
semiconductor, 9
sensing, 2

sensitivity, 8, 69
sensors, vii, 1, 2, 4, 5, 7, 8, 9, 10, 11, 25, 28, 32, 41
services, 22, 23, 24
signals, 12
signs, 60
simulation, 20, 21, 39, 64, 86, 89, 91, 101, 107
software, 2, 4, 5, 12, 14, 15, 17, 18, 20, 21, 22, 23, 24, 25, 28, 29, 37
SPEAR earpiece (figure 1), 86
specifications, 11, 32
speech, vii, viii, 65, 83, 84, 85, 86, 87, 88, 89, 90, 94, 99, 100, 101, 102, 103, 104, 124
speech command recognition (SCR), 87
state, 4, 13, 14,18, 25, 32, 60, 102
statistics, 44, 48, 90
stress, 54, 85, 102
structure, 22, 24
synchronization, 19, 20, 21

T

target, 18, 21, 54, 84, 86, 87, 99, 101, 103
task demands, 33, 54
task load, 56
task performance, 33, 34, 48, 84, 103
team members, 85
technology, vii, 1, 2, 3, 6, 8, 25, 86
teleoperation tasks, viii, 32
testing, 20, 21
Think-A¬Move (TAM), 86
time pressure, 54

topology, 19
trade, 28, 55, 104
training, 34, 40, 41, 45, 64, 65, 66, 67, 84, 85, 88, 89, 90, 91, 97, 98, 99, 100, 101

U

United States (USA), iv, 8, 39
unmanned ground vehicle (UGV), 103
Unmanned Vehicle Technologies Division (UVTD), vii, 2
urban, vii, 1, 37, 67, 82

V

vehicles, 7, 18, 40, 64, 89, 107
velocity, 12, 16
vibration, 4, 85
video games, 64, 91, 107
vision, vii, 1, 5, 9, 16, 17, 26, 58, 63, 66, 70, 71, 106
vocabulary, 85

W

walking, 27, 84
war, 39, 82
Washington, 56, 104
weapons, 38, 39, 59, 66
wear, 63, 106
workload, 9, 33, 39, 43, 49, 51, 54
workstation, 19, 20, 21